Lost Treasures
&
Old Mines

A New Mexico Federal Writers' Project Book

Lost Treasures & Old Mines

A New Mexico Federal Writers' Project Book

Compiled and Edited
by
Ann Lacy
and
Anne Valley-Fox

SANTA FE

© 2011 by Ann Lacy and Anne Valley-Fox.
All Rights Reserved.

No part of this book may be reproduced in any form or by any electronic or mechanical means including information storage and retrieval systems without permission in writing from the publisher, except by a reviewer who may quote brief passages in a review.

Sunstone books may be purchased for educational, business, or sales promotional use. For information please write: Special Markets Department, Sunstone Press, P.O. Box 2321, Santa Fe, New Mexico 87504-2321.

Book and Cover design › Vicki Ahl
Body typeface › Bell MT
Printed on acid free paper

Library of Congress Cataloging-in-Publication Data

Lacy, Ann, 1945-
  Lost treasures & old mines : a New Mexico Federal Writers' project book / compiled and edited by Ann Lacy and Anne Valley-Fox.
    p. cm.
  Includes bibliographical references.
  ISBN 978-0-86534-820-2 (softcover : alk. paper)
 1. New Mexico—History, Local—Anecdotes. 2. Mines and mineral resources—New Mexico—History—Anecdotes. 3. Frontier and pioneer life—New Mexico—Anecdotes. 4. Pioneers—New Mexico—Biography—Anecdotes. 5. New Mexico—Social life and customs—Anecdotes. 6. New Mexico—Biography—Anecdotes. I. Fox, Anne Valley. II. Federal Writers' Project. New Mexico. III. Title. IV. Title: Lost treasures and old mines.
  F796.6.L33 2011
  978.9—dc23
                              2011029103

WWW.SUNSTONEPRESS.COM
SUNSTONE PRESS / POST OFFICE BOX 2321 / SANTA FE, NM 87504-2321 /USA
(505) 988-4418 / ORDERS ONLY (800) 243-5644 / FAX (505) 988-1025

Dedicated to the writers of the
New Mexico Federal Writers'
Project, 1935–1943.

CONTENTS

Acknowledgements ~ 11
Editors' Preface ~ 13
About the New Mexico Federal Writers' Project ~ 15
Foreword by Jon Hunner, PhD ~ 17
Map of the Territory of New Mexico, ca. 1857 ~ 21

PART ONE: LOST TREASURES

Gold Fever in Ojo de la Casa as Told by Patricio Gallegos by J. P. Batchen ~ 25
Out of Bondage (José de Luz Seeks His Fortune) by Lou Sage Batchen ~28
The Adams Diggings by F. V. Batchler ~ 33
Lost Treasure by Manuel Berg ~ 40
Chilili by Lorin W. Brown ~ 43
Gambusinos by Lorin W. Brown ~ 44
Secrets of the Guadalupes by Lorin Brown ~ 47
The Lost Mondragon Mine by Lorin W. Brown ~ 49
Buried Treasure by Genevieve Chapin ~ 54
Lost Treasures of Grand Quivira by Edith Crawford ~ 56
Tecolote by Lester Raines ~ 57
Prospector 40 Years by Edith L. Crawford ~ 58
The Lost Gold Mine by Kenneth Fordyce ~ 60
Treasure by Ramitos Montoya ~ 62
An Expedition from New Mexico Writers' Project ~ 64
The Haunted House by Lester Raines ~ 65
The Lost Sublett Mine by Katherine Ragsdale ~ 66
Pioneer: Hidden Treasure (Cutter, Sierra County) by Lester Raines ~ 68
Still Buried Treasure by Hermione Manning by L. Raines ~ 69
The Church of the Golden Bell by L. Raines ~ 72
The Story of Adam's Diggings by L. Raines ~ 73
The Treasure of Punta de Agua by Edna Shaw by L. Raines ~ 75
Buried Treasure by Lester Raines ~ 77
Indian Fight in the Floridas by Betty Reich ~ 78
Indian Stagecoach Robbery by Betty Reich ~ 80

The Lost Mine by Betty Reich ~ 81
The Schaeffer Diggings by Betty Reich ~ 82
Jessie Martin, Desert Rat by N. Howard Thorp ~ 84
La Mina Escondida: The Hidden Mine by N. Howard Thorp ~ 88
Lost Mine of the Pedernal by N. Howard Thorp ~ 92
The Dead Burro Mine by N. Howard Thorp ~ 96
A Discovery of a Cave by Mrs. Frances Totty ~ 100
Buried Money on the Mimbres by Frances E. Totty ~101
Hidden Treasures by Frances Totty ~ 103
The Story by F. Totty ~ 105
Louis Ancheta by Frances E. Totty ~ 106
The Ghost of Georgetown by Mrs. W. Totty ~ 108
Lost Treasures by Mrs. W. C. Totty ~ 110
The Ghost of Priors Canyon by Frances E. Totty ~ 111
Wilcox Mining Claim by Frances E. Totty ~ 114

PART TWO: OLD MINES

Humorous Incidents of Early Mining Days: A False Alarm by James A. Burns ~ 141
Humorous Incidents of Early Mining Days: A Self Made Reputation by James A. Burns ~ 143
First Mine Registration in New Mexico, 1685, Pedro de Abalos ~ 146
Mines of Northern New Mexico (Historical) by James Burns ~ 147
Place Names: Cities, Towns and Villages, Lincoln County by Edith L. Crawford ~ 150
The North and South Homestake Mines by Edith L. Crawford ~ 151
The Old Abe Mine (I) by Edith L. Crawford ~ 153
The Old Abe Mine (II) by Edith L. Crawford ~ 156
Tin Pan Canyon, Colfax County by Kenneth Fordyce ~ 158
Coal in Colfax County—New Mexico by Kenneth Fordyce ~ 159
Elizabethtown by Kenneth Fordyce ~ 161
Description of a Mine: Santa Rita Copper Mine by Mrs. Mildred Jordan ~ 164
Description of a Point of Interest: Steeple Rock Peak by Mrs. Mildred Jordan ~ 167
Sad Disaster at Mogollon by John Looney ~ 168
Citizenship Papers: Celestino Carrillo by Ernest Prescott Morey ~ 170
Glossary of Mining Idioms Used in New Mexico by T. F. Bledsoe ~ 174
Reminiscence of an Old Prospector by Ernest Prescott Morey ~ 176

Tyrone Turquoise Mines by Ernest Prescott Morey ~ 183

Hands That Built America: Memorandum On Mining Operations from New Mexico Federal Writers' Project ~ 187

Mining Stories from Las Placitas: Legend of Montezuma Mine from New Mexico Federal Writers' Project ~ 191

A Mine for Two Barrels of Water by W. L. Patterson ~ 192

The Helen Rae Mine by W. L. Patterson ~ 194

Turquoise Mines Near Oro Grande by W. L. Patterson ~ 195

Mina de la Tierra by Robert Pfanner ~ 197

Mining and Minerals by Harriet Brent, rewritten by Robert Pfanner ~ 198

Letter from Jeff Boone to Mr. Charles Ethridge Minton ~ 200

Gold by D. D. Sharp ~ 201

La Mina de la Virgin de Oro or The Mine of the Gold Virgin by N. Howard Thorp ~ 205

Dolores from New Mexico Federal Writers' Project ~ 209

The Old Mines Near Las Placitas by N. Howard Thorp ~ 210

A Prospector's Experience by Mrs. W. C. Totty ~ 213

Founding of Silver City by Mrs. W. C. Totty ~ 217

Gold Gulch Findings by Frances E. Totty ~ 219

Mining Life by Frances E. Totty ~ 220

Negro Findings by Frances E. Totty ~ 222

Notes on Pinos Altos Range by Mrs. Mildred Jordan ~ 223

Pinos Altos: The Great Gold Producing Camp of the County by Frances E. Totty ~ 224

Silver City Mines by Frances E. Totty ~ 227

Snakes in a Mine Shaft by Mrs. W. C. Totty ~ 231

Description of an Old Ghost Town: Georgetown by Mrs. Mildred Jordan ~ 232

The Founding of Silver City, Taken from the Mogollon Mines, Published in 1914 by Frances E. Totty ~ 233

Turquoise by Mrs. W. C. Totty ~ 236

Kingston, Ghost Mining Town by Clay W. Vaden ~ 241

Old Rifle Bears Marks of Man's Fight With Bear by Clay W. Vaden ~ 243

Ox Team Freighter Recalls Old Days in Kingston Mine Area by Clay W. Vaden ~ 244

List of Facsimiles ~ 247
List of Illustrations ~ 248
Bibliography of New Mexico Federal Writers' Project Documents ~ 252
Names Index ~ 259

Acknowledgments

We wish to thank the New Mexico State Records Center and Archives, the Museum of New Mexico Palace of the Governors Photo Archives and the Fray Angélico Chávez History Library, Santa Fe, New Mexico for the use of their collections.

We are grateful to the archivists at the NMSRCA for their able assistance with our research.

Special thanks to Project Crossroads and Elise Rymer for her inspiration and enduring support.

Editors' Preface

While researching New Deal records at the New Mexico State Records Center and Archives in Santa Fe, New Mexico, we discovered a treasure trove of folders labeled "WPA 1936-1939." Inside were hundreds of manuscripts pecked out on old upright typewriters by New Mexico writers determined to make a buck by their wits while documenting some of the state's historical highlights. The first book, *Outlaws & Desperados: A New Mexico Federal Writers' Project Book*, was published in 2008 to mark the 75th anniversary of the New Deal; the second, *Frontier Stories: A New Mexico Federal Writers' Project Book*, followed in 2010. This volume, *Lost Treasures & Old Mines*, is the third in the series.

Between 1936 and 1940, the writers from the New Deal's New Mexico Federal Writers' Project (NMFWP) collected stories throughout New Mexico describing a time that was beginning to fade into history. The experiences and exploits of settlers and earlier inhabitants of the New Mexico Territory during the territorial days after 1846 gave way to a less isolated and more modern era beginning with statehood in 1912. By the 1930s, NMFWP writers were recording the stories of old-timers reflecting back on New Mexico's vanishing past.

The stories in this volume offer many colorful first-hand accounts of life on the frontier. It is important to remember, however, that not all perspectives are represented in the WPA archives. The voices of Native Americans, for instance, rarely show up in the New Mexico Writers' Project interviews, and first-person narratives by women are also rare. Therefore, in the words of former state historian Estévan Rael-Gálvez, readers should be encouraged to "read between the lines."

With a view towards authenticity, the writers of the New Mexico Federal Writers' Project attempted to capture each informant's particular way of speaking; the oral histories in this collection reflect some of the colloquialisms of old Territorial days. As editors, we have tried to stay close to the original manuscripts and have corrected punctuation and spelling only when necessary for readability and clarity. For the most part, the manuscripts stand close to their original archival versions. We hope you enjoy *Lost Treasures & Old Mines* as a rich expression of New Mexico's colorful past.

—Ann Lacy and Anne Valley-Fox
Santa Fe, New Mexico

About the New Mexico Federal Writers' Project

The Great Depression that came on the heels of the stock market crash of 1929 threw the country's financial institutions into chaos and put many people across the nation out of work. In 1933, President Franklin Delano Roosevelt inaugurated his New Deal administration, a comprehensive program designed to stimulate the country's economy while lending a hand to the unemployed. March, 2008, marked the seventy-fifth anniversary of the New Deal.

At a time when many people were down on their luck during the Great Depression, the New Deal's New Mexico Federal Writers' Project (NMFWP) employed writers around the state to record the extraordinary history and lore of New Mexico. The Federal Writers' Project was one of a number of white-collar relief projects of the Works Progress Administration (WPA) that put Americans back to work. In addition to the Federal Writers' Project (FWP), the projects included the Federal Art Project, the Federal Music Project, the Federal Theater Project and the Historical Records Survey.

The New Mexico Federal Writers' Project was officially launched on August 2, 1935, under the direction of poet and writer Ina Sizer Cassidy. Between October, 1935, and August, 1939, a cadre of field writers wrote stories, collected articles, conducted interviews and transposed documents for the public record. Although each of the 48 states across the nation launched their own Federal Writers' Project, New Mexico was seen as geographically and culturally unique. From his office in Washington, DC, the national director of the Federal Writers' Project, Henry G. Alsberg, urged New Mexico project writers to emphasize the state's visual, scenic and human interest subjects in the project's guide, *New Mexico: A Guide to the Colorful State*. "Try to make the readers see the white midsummer haze, the dust that rises in unpaved New Mexican streets, the slithery red earth roads of winter, the purple shadows of later afternoon . . . ," he told them.

New Mexico field writers apparently felt a similar enthusiasm, as they created hundreds of documents to preserve the state's vivid lore, scenic locale and colorful past for future generations. Their subjects ranged from the colonial New Mexico days of the 1600s and 1700s to the beginnings of the 1900s—from horse-drawn cart to car. Their many lively selections included firsthand oral accounts and remembrances by settlers and residents who lived to tell the story of New Mexico's Territorial days.

In 1939, under the WPA's reorganization, the New Mexico Federal Writers' Project became the Writers' Program. By that time, Aileen O'Bryan Nussbaum had replaced Ina Sizer Cassidy as project director. In Washington, DC, Charles Ethrige Minton supervised the New Mexico Writers Program until its closure in 1943. Through its tenure, the New Mexico program produced *Calendar of Events*, written by project writers and illustrated by Federal New Mexico Art Project artists, as well as *Over the Turquoise Trail* and *The Turquoise Trail*, two anthologies of New Mexican poems, stories, and folklore. A major achievement of the FWP was an American Guide Series publication entitled *New Mexico: A Guide to the Colorful State*, first published in 1940.

Project writers in New Mexico had a wealth of sources to draw upon and they mined them well. They collected tales from old-timers with a colorful heritage and culture; there were early explorers, diarists and journalists, poets and artists, miners, ranchers and cowboys, farmers and merchants, lawmen and outlaws, anthropologists and folklorists. These are the voices of the many travelers—paso por aquí—who animate New Mexico history.

The efforts of the NMFWP field workers have left us a rich compilation of documents stored in various collections in New Mexico, including the New Mexico State Archives as well as museum and university collections. The Library of Congress in Washington, DC also holds copies of many of the manuscripts. Now, seventy-eight years after FDR launched the New Deal, with the New Mexico Federal Writers' Project book series a substantial number of these readings have found their way out of archival folders and into print for the public's enjoyment.

Foreword

by Jon Hunner, PhD

Few things stir our blood as much as the hint of a lost treasure, especially if gold is involved. *Lost Treasures & Old Mines* targets this perennially popular topic for those of us who are armchair vagabonds. The lure of gold, either lost or found, captures our imagination as we dream about the riches that can change people's lives overnight. The stories in this volume about mining run the gamut from gold and other precious metals to coal and from grizzled prospectors scratching in desert hills to Native Americans guarding sacred caves full of wealth. *Lost Treasures and Old Mines*, ably complied by Ann Lacy and Anne Valley-Fox, has something for anyone who is fascinated by humans who have struck it rich, lost it all, hidden it away, or have searched for someone else's misplaced treasure. There is, as the saying somewhat goes, gold in these thar pages.

The stories in *Lost Treasures & Old Mines* have been mined from the mother lode at the New Mexico State Archives and Records Center. Throughout the 1930s, twenty-four researchers in New Mexico canvassed the state, writing down some 218 stories that people told them about a wide range of topics. In bringing these Federal Writers' Project stories to light, the editors have uncovered a true lost treasure.

These tales descend into the cavernous shafts of our region's history. Many of the stories focus on the 19th century, when intense mining activity occurred in New Mexico; however, others dig deeper and recount Spanish Colonial legends about lost treasures. Some accounts go even further back to pre-contact times and retell tales passed down through generations of Native Americans. Before the Europeans arrived, turquoise was the precious stone coveted by humans, and there are several stories here about such ancient mines.

The pursuit of wealth by digging into the ground brings out the best, but also the worst in us. Jumping someone else's claim, hiding a rich strike, stealing a season's stake, even killing a partner, all are part of the fascinating stories in this book about the conflicts, the sacrifices, the triumphs, and the hardships of lives impacted by mines and treasures. The nuggets of gold found in this volume shine without a lot of polishing.

Some of the gems revealed include a story written down by Lorin Brown about Juan Mondragon, a young sheepherder near Mora, who refused to follow and spy on two miners when they returned to their strike in the mountains because, "Quien sabe, one does not get more than he is supposed to in this world." Edith Crawford related a story about Grand Quivira, rumored to be the burial site of treasure during the Pueblo Revolt in 1680. From a 1899 White Oaks Eagle newspaper article, this account about the retreat from Santa Fe captures our attention: "We traveled southward, fighting all the way. We have 270 burro loads of treasure. We buried it between two little hills, burning brush over the spot." In the 1930s, holes still abounded in the hills around Grand Quivira from people hunting for the 250 year old treasure. In southeast New Mexico, Frances Totty heard the story about the Ancheta family, who for several generations suffered the curse of acquiring a large fortune but not being able to enjoy the pleasures "that money could give." Next door neighbor Pedro Raesequeon even felt the curse as he buried $17,000 dollars on his land and later could not find the cache. Donaciano Gallegos told FWP researcher N. Howard Thorp about La Mina de la Virgen de Oro (The Mine of the Golden Virgin) on the western slopes of the Sandia Mountains. As a young boy in the mid 19th century, Donaciano herded goats in the foothills of the Sandias. Following his goats who had disappeared into a cave, he stumbled upon an image of the Virgin made out of gold sitting on a shelf. Because the cave was on Pueblo of Sandia land, three Native Americans ran him off, and even though he returned with friends later on, he was never able to relocate the cave. Clay W. Vaden interviewed Cebe Goins, who at 90 years old, told Clay about one of the richest silver strikes in New Mexico. The Bridal Chamber Mine in Lake Valley, northwest of Hatch, had blocks of pure silver measuring one square yard removed. In 1885, almost $7,000,000 of silver came out of Lake Valley, which in 2008 would amount to about $165,000,000.

These stories in *Lost Treasures & Old Mines* offer many accounts about mining and treasures in New Mexico, from gold to coal, and other tales, sometimes quite tall, about wealth lost and found. Additionally, there are many accounts of life in New Mexico, from Spanish colonial times to the first part of the 20th century.

Some lost treasures in this book have nothing to do with precious metal. The forgotten folklore and heritage that these stories uncover provide rich details about life in New Mexico. In Pinos Altos just north of Silver City, after gold was discovered in 1860, 700 men congregated under the Ponderosa Pines and created an instant city. The Civil War played a part in this gold rush as some miners left the Tall Pines to fight in the Union or Confederate armies. Conflict between miners and Native Americans also figure in some of the stories as well as stories about prospectors who served as cultural brokers between the indigenous peoples who claimed the ore-rich areas as their ancestral lands and those who arrived with a pickax and a shovel to disrupt the countryside. As a prelude to the story about the Golden Virgin, a history of the Pueblo of Sandia describes its long and deep history during the Spanish colonial period. At other times, we hear about subsistence farmers, barely able to make ends meet, who stumble onto a windfall of treasure in lost or found gold. Granted, the main focus of *Lost Treasures & Old Mines* centers on mining and lost wealth, but the stories also convey a sense of how people lived in the past and evoke the isolation that farmers, ranchers, and prospectors all felt in the remote mountains and deserts of New Mexico.

The history of mining in New Mexico as elsewhere resounds with stories of booms and busts. Instant towns erupt overnight on sides of mountains or in dry arroyos, and then once the lode plays out, everyone chases after the next strike, abandoning their homes and businesses. Although these places hold little worth without the precious metals, they still attract our interest. For those who want to visit these ghost towns and walk in the hobnailed footsteps of miners, camp followers, and ne'er do wells, some of these stories will introduce you to new places for exploration. Using your historical imagination, you can fill the dusty, lonely streets and ruins with the noises of the stamping of the mill, the shouts of discovery, the crack of gunfire. These old mining settlements dot our landscape, like nuggets of placer gold scattered around the countryside of New Mexico.

Lost Treasures & Old Mines assembles numerous stories about a vital part of the region's history. Taken from the accounts preserved by the Federal Writers Project, including articles published in frontier newspapers, these tales revolve around the wealth that we find and lose, but also are gems in their own rights. Lacy and Valley-Fox have mined these rich veins to produce a fascinating and entertaining book.

—Jon Hunner, PhD
Mesilla, New Mexico

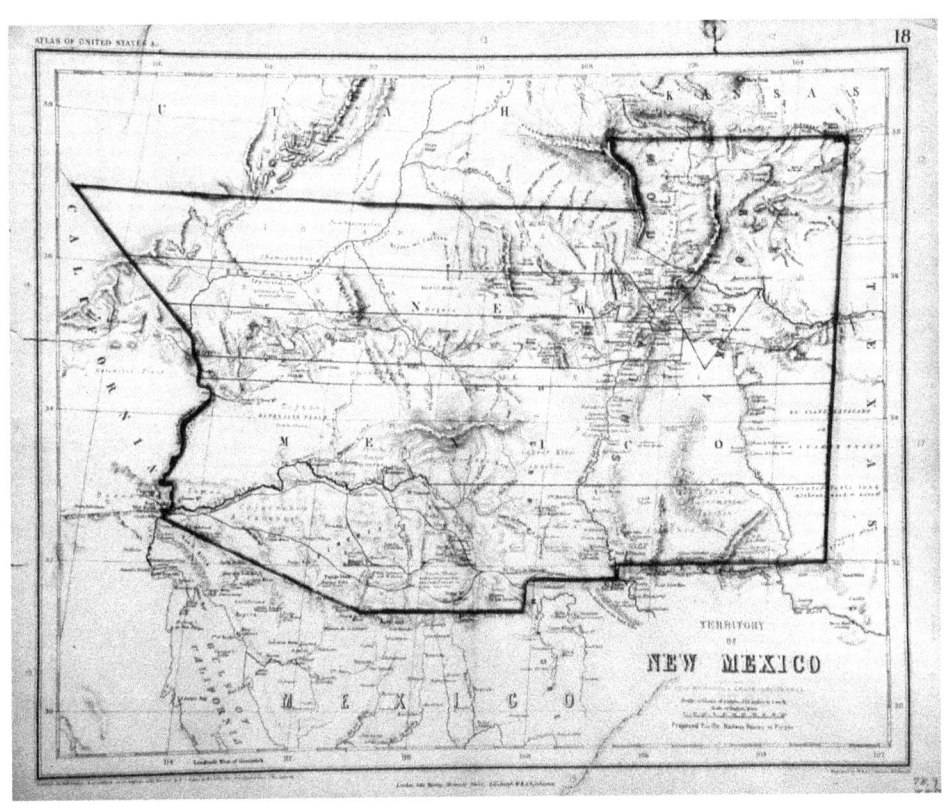

A. D. Rogers, A. Keith Johnston, Territory of New Mexico, Fray Angélico Chávez History Library, Map Collection (78.9) ca. 1857

PART ONE: LOST TREASURES

"There have been many stories told of hidden gold, but few of them have any foundation in fact."
—From "Indian Stage Coach Robbery"
by Betty Reich

Gold Fever in Ojo de la Casa as Told by Patricio Gallegos

by J. P. Batchen

Everywhere men gathered along the old Santa Fe Trail and told of tales they had heard of New Mexico, there was repeated by someone the legend of the old Montezuma Mine located somewhere at the north end of the Sandia Mountains. Dick Wooton heard of it at Trinidad and at Raton Pass. He became interested to the point of trying to locate it. As the legend goes, every sign of the locations was wiped out. He could find nothing. But he did find an old Indian in the vicinity who lay sick on the floor of his little house. No one lived near him and he was too ill to rise to his feet. Wooton gave him some of the food he had brought with him. He nursed the old Indian back to health. Then he asked him where was the Montezuma Mine. The story goes that it took Wooton several days to persuade the old Indian to talk; then he was told that he could not tell anyone of the spot. Wooton waited and continued sharing his grubstake. Finally the old Indian confided to Wooton that he was most grateful for the care he had given him and he felt that Wooton had saved his life and while he could not tell him, he told Wooton to watch him as he walked along the mountain side and note the places he stopped. And so it was that later Wooton discovered arrows cut in the rocks pointing in one direction. He located a spot and set to work with his pick and shovel. He did unearth a working that had been filled in, or else had caved in. For years the mine, or rather, workings that resulted, was called "The Wooton Mine."

But it was in the late 1870s and the early eighties that things happened in Ojo de la Casa to break the monotony of goat herding, wood hauling, and food problems. The young prospector who had somehow possessed himself of the house of Santos Lovato started a hunt for gold, egged on by the tempting legend of the old Montezuma Mine. And the hunt was stimulated by the fact that there was an awakening in New Mexico to the possibilities of hidden mineral wealth in the area from Santa Fe to the south end of the Sandias. On the high tide of the excitement, a prospector named Wilson came into Ojo de la Casa. He located claims at the north end of the Sandias where legend located the old Montezuma. Many prospectors followed him and soon the whole Sandia ridge, the length of Las Huertas Cañon, was staked with claims and the locations registered at Albuquerque, the county seat.

All this meant assessment work if the claims were to be held according to the law. Ojo de la Casa was alive to the earning power of the pick and shovel in their hands. Digging started, times prospered, and everyone became accustomed to the chink of silver in his pocket. Wilson opened a saloon, and the inevitable gaming table of the day went with it. The peace and quiet of the hidden village was lost in the din of blasts of gunpowder, and the drunken brawls originating at Wilson's saloon and terminating anywhere and everywhere. The sudden and mysterious slaying of Wilson and the looting of his place closed the saloon. He was buried with scant ceremony in his door yard, for he had a highly developed ability for piling up ill will against himself. The murder never came to the eye of the law, and so it passed. By legal observances his mine came into the possession of the young prospector. But by now the gold fever had burned itself out in Ojo de la Casa. Besides, the Bland gold fever was on.

But before Ojo de la Casa settled back into lethargy, old Francisco staged a last gold hunt. He thought backward and forward, dipped his hands into his empty pockets, then he sought the prospector, who was not so young anymore but who still believed in tales of hidden treasure. Francisco told him a tale of his youth, how one day while he was herding his goats in the mountains one kid strayed from the flock and became lost. He hunted everywhere up and down the mountainside until he himself became lost. Then he heard the cry of the kid in a thick underbrush. Oaks so thick that he could not see through them, so he crawled along looking, breaking his way with his hands. Suddenly he came to the edge of a deep cave with a ladder sticking out of it. He wanted to see what was in the cave, so he went down the ladder and there at the bottom of the cave he saw a little room dug out of one side. In that room he saw piles of shining gold and silver bars. He was so frightened at seeing so much treasure right there before him that he did not hear anyone come down the ladder. The first thing he knew, somebody threw a blanket over his head and he was carried up the ladder and a long way off from the cave. Then rough hands stood him on the ground and jerked the blanket from his head and then he saw angry Indians looking at him. They told him if he ever came near that cave again they would kill him. So none of them ever tried to find it.

The prospector hired him at a dollar a day and his dinner to take him over the ground where the cave was located. Francisco's old eyes lighted, and for a week they scoured the vicinity where the cave was hidden in the thick brush, but no evidence of it could they find. When old Francisco sensed that the prospector was ready to give up, he said, "That be a long time ago. Big rains he come every year. He fill up holes, he make big change—now—no can find."

Source of Information: Stories of William Eckert and Juan Maria Gallegos.

Out of Bondage
(José de Luz Seeks His Fortune)

by Lou Sage Batchen

Peonage was a recognized institution in the Spanish Colonies of the Southwest. There were laws giving recognition to both patron and peon; though it never got further than the books in which such rights, privileges and obligations were written, there was a definite obligation on the part of the patron to the peon. Whatever the peon bought at the store of the patron, he was to pay the market price for it. Then there was another: only two thirds of the peon's wage per month could be credited to his accounts. The other one third, or two dollars sixty-six cents (if the peon received eight dollars per month) or, one dollar sixty-six cents (should his monthly wage be five dollars) was to be paid each month to the peon. Neither of these particular provisions were observed. The prices charged at the patron's store for every item bought at the store were enormous. This kept the poor peon in debt. There was no way he could ever pay out—and still live. There was no set amount, if any, paid the peon in cash, as part of his monthly wage. But there was one provision of the law which applied to the peon which was kept to the letter. The patron saw to it. The peon could bind his children to his patron to work out his debts to the patron. This, the peon was forced to do.

This custom prevailed so long in the Southwest that the peons had few if any dreams of life beyond that of being bound to some great Family who would keep them from actually starving to death. However, the patrons were in no way obligated to see that their peons did not starve. It was to their advantage to see that they did not. This tiny hold on social security kept the peon laboring on, binding his sons to the same sort of life he led himself, with no time of their own, or means with which to do anything for themselves. Such was the status of the lowly peon when he found himself suddenly cut adrift from his patron in 1867 when the system of peonage was abolished. Or, that was, the peon would be released as soon as his indebtedness was cancelled. That brought about an immediate change. Less credit, that debts could the sooner be wiped out.

All peons were not affected equally. As men are different, so are they differently moved by the same circumstances. When some of the peons realized that what little security they had was taken from them, by a law they had not sought, they were paralyzed with fear for the future. They stood humbly before their patron and cried to be "occupied" again. Others, fearful of being upon their own, took refuge with their families, in the pueblos with the Indians where they had blood ties. Yet others calmly accepted their fate, turned their faces steadily in a new direction of living for themselves, sought to clear and cultivate new land and acquire flocks of their own. There were yet others who, while in the occasional service and debt of the patron, were still free enough to make most of their living for themselves independent of him. Most of those in Las Placitas belonged to these last two groups. They could yet find occupation with Don José Leander Perea, receive credit and take off time to follow their personal pursuits. These were the sort of men who did find employment with the patron after the abolition of peonage and credit.

But while the former peons were struggling for readjustment, there came a great new adventure in the way of labor. The railroad was coming! Some of the huskier, more ambitious ones found work, work which paid cash for labor done: hard labor but wonderful wages to the peon who had given all he had in the way of strength and loyalty for five and eight dollars a month.

But there was another and greater adventure, from the new freed man's point of view, which brought work and money right into his midst. It was the rush and the excitement of the mining boom, the hunt for gold. He could help the gringos hunt rocks and dig holes into the sides of the mountains and get paid for it. The gringos knew where there was gold to be found. In the mountains, but they must find the right spots and dig deep enough and in the right direction. Lots of digging to do.

Some of the men worked for the gringos at Las Placitas where much assessment work was done. One of those men was José de Luz Montoya, once peon in the service of Don José Leander Perea, who reached his year of freedom about three years after the law of 1867, as it took him that time to work out his obligation and keep body and soul of his family together. José de Luz was one of those, too, who, when separated from the little he had of security, was full of fear for the future. He knew not which way to turn. With but a house and a small plot of ground of his own, and with no blood ties in any of the pueblos, his outlook was bleak indeed. When he found that his once patron would not employ him further he went to work as many others did, cleared and planted a few fields and hunted wild game. Somehow his family carried on, even though José de Luz lost rather than accumulated ambition as he went along.

Then came a gringo to Ojo de la Casa. He knew he would find gold in the vicinity of the old Montezuma mine. Old legends told of gold being found there. He hired José de Luz to help him find it. José de Luz delved into the mountainside, and received one silver dollar for each day he delved. His wife fixed up their old adobe house and he spent silver dollars to buy desirable things with which to fill up their stomachs to capacity. A luxury hitherto denied them.

But this new and costly standard of living could be maintained only by some daily ten hours of delving by José de Luz. He was wholly unaccustomed to such strenuous labor. In service to his patron he herded sheep. It was a leisurely existence compared with handling a heavy pick and shovel all day long. He cast about for a change of employment. By this time he felt that the prospector, the gringo who knew that he could find gold in the region of the Montezuma, had told or taught him all there was to know about the business of finding gold.

So it was that José de Luz decided to find gold for himself. And as a sideline, he accumulated a small herd of goats, which he would herd about as he scoured the arroyos for the right kind of rocks and thoroughly investigated every hole or cave he came upon. There would be excellent pasturage for the goats: they would grow fat and make fine eating, or fetch a good price if he chose to sell them. The only real trouble he would have would be keeping clear of bears. While herding his patron's sheep he had learned much about bears; mainly to keep clear of them.

Fate seemed to favor José de Luz in the matter of bears and mountain lions. He did keep out of their way. Farther and farther he wandered from the old herding ground at Cerro Pelon. He explored each cave or mine working he passed; he gathered samples of rock with a gleaming yellow streak in them. His dreams of finding gold would surely come true.

One day José de Luz lost some of his goats. He started to track them down. He entered strange territory as he climbed the higher mountains and descended slopes he had never known existed. Grass grew rank there and he saw goats feeding in the distance. His lost goats of course. He set out to herd them back.

On his way he spied a cave. He approached it. It went far down into the earth and there was a ladder leading down into it. Without debating the matter in his mind, José de Luz set his foot on the top crosspiece and hastily descended into the cave.

But he was to be frightened half out of his wits before he had been there enough to make it all out. Swiftly, silently, something surrounded him, covered his head with a blanket, seized him, and dragged him back up the ladder. But that was not all. He was rushed so fast over the ground that he could scarcely keep his feet.

He did know into whose hands he had fallen. He fearfully, suddenly, knew the reason. At once he knew he had little chance of his life. He was in the clutches of the Sandia Indians. Unwittingly, he had invaded their private domain.

"What did you see down in that cave?" demanded an angry voice right in his ear.

"Nothing, nothing," protested José de Luz. "I did not have the time. It was so dark my eyes could not see."

The Indians demanded over and over what he had seen while in the cave. Over and over their prisoner declared that he had seen nothing. Handle him as roughly as they pleased, they could not shake his statement from the original words he spoke to them.

They came to a halt. Angry voices sounded all about them. The cover was jerked from him as suddenly as it had been flung over him. He was completely surrounded by angry, menacing Indians. Some shouting, "Kill him! Kill him!"

What chance had he? Trembling violently he protested loudly, "I saw nothing. I could not. It was dark. I was there too short a time. My eyes could not see!" But the angry voices about him continued to say, "Kill him!"

Then José de Luz saw one friendly face. It was his compadre. This Indian was the padrino (godfather) of the first son of José de Luz. He looked at José de Luz sympathetically, and the poor frightened man saw a ray of hope. When the Indian asked him whether he had seen anything in the cave, he stuck to the story he had told the other Indians. At that, the friendly Indian merely said to his companions, "I say, let him go."

Reluctantly José de Luz was released. But he was not permitted to go for his goats. The grateful man was not much troubled by the refusal. He was happy to make a hasty departure from his would-be killers and he lost not one moment in covering the ground between the pueblo of Sandia and his home at Las Placitas.

José de Luz forthwith lost his taste for investigating caves, or even hunting rocks streaked with shining yellow. He accumulated more goats and more fields to clear and till. He became a quiet, stay-at-home man.

Not until after the death of José de Luz was the story of the adventure repeated outside his immediate family. Fear of the consequences sealed their lips; for he did see something of what was in the cave. There was an altar adorned with beautiful images, small ones there were. And other bright colored things he had not the time to make out.

To this day, if the story is repeated, it is spoken cautiously. And the reason? It is believed that it is the cave which holds the treasures hidden there by the Indians, when the Spaniards were driven out. It is guarded to this day. Anyone entering that vicinity is watched by the Indians with hostile eyes. None from Las Placitas would go there.

There is scarcely a native in Las Placitas but believes that the tragedy which happened to the two American boys, in that vicinity a very few years ago, was not an accident. They were too near the treasure. The treasure is always guarded by zealous and hostile Indians. Both boys were found dead in that vicinity.

Source of Information: José Gurule, age 90, Las Placitas, New Mexico, who knew the José de Luz of this story, as well as the conditions which existed after the abolition of peonage.

The Adams Diggings

by E. V. Batchler

Since I came to New Mexico, eighteen years ago, I have heard stories of the wealth of the famous, old, lost Adams Diggings Mine. I have heard at least a dozen different stories and each succeeding story made the mine richer both in actual gold value and romantic interest. As is often the way with lost mines of this type, it all depends on who you listen to, whether the mine gets richer or not. It always seemed strange to me that nearly every old-timer will swear that he knows more about a fabulously rich, lost mine than any other old prospector. He will try to discredit other prospectors who have searched for the mine and in an effort to tell something "bigger," magnify its riches by manyfold what others have estimated it at. In reality, none of them know or have the slightest idea as to the value of the lost mine, because it has never been found.

The current story and the one that seems to be the most popular, is one that I read in the El Paso Herald a few years ago. It stated that a bunch of men, among them Edward Adams, who purportedly found the mine that was later named for him, organized an expedition to go to California. Their probable starting place was Magdalena. They traveled in a northwesterly direction, until somewhere between Magdalena and Old Fort San Rafael, they camped on a little stream.

One of the men noticed gold in the stream and excitedly revealed his discovery to the rest. Adams, who knew a little more about mining than his companions, decided that the gold washed into the stream was from a rich outcropping above the camp. Taking his partner, a man by the name of Davidson with him, he left camp and traveled up the canyon about a mile to try to discover the "mother lode."

A little while after they had disappeared around a bend in the creek, the expedition was attacked by Apaches, and as they caught the encampment totally unprepared, the Indians massacred every man in camp.

Adams and Davidson heard the firing, and suspecting its cause, took to the cover of the bushes on the nearby hillside. After hiding for several hours, the two men cautiously made their way over the hill and saw that the Apaches had left, secure in the belief that they had killed all the men of the expedition, and had taken all the mules and horses with them.

After burying all the dead, Adams and Davidson knocked a few pieces of gold-bearing ore off an outcropping of quartz that they believed to be the "mother lode." They then purportedly made their way to Fort San Rafael, where they said they asked for aid to go back and find the gold and were refused by the officer in charge.

They then made their way afoot and after perilous hardships and a great deal of suffering, came into the little town of Reserve, in what is now Catron County. It is said that they showed samples of the ore to several of the natives, and then after borrowing some money on the strength of the richness of the ore, bought horses and went to Pima, Arizona, where Adams had friends whom he thought had enough money to properly outfit an expedition to return to the place where he had found the gold.

The expedition was organized, and traveled from Pima to Alma and thence to the immediate locality where Adams was supposed to have found the gold. But through some freak of nature or loss of direction, they could not find the gold, or even the place where the men had been massacred. Perhaps it was because Adams and Davidson both were notoriously poor in remembering directions. Many expeditions have been organized since then, but to this day, the Adams Diggings remains as much a mystery as when Adams first told of it.

Now I am going to tell a story that is almost completely at variance with the story printed by the El Paso Herald. It is a firsthand story from the lips of Bob Lewis, pioneer, old-time prospector, cowboy, and for the better part of his manhood, a frontier peace officer and a personal friend of Edward Adams. Bob is a big man, well over six-feet and weighing in the vicinity of two-hundred pounds. He always has a jovial greeting and manner, and has the map of Ireland printed all over his face. Big, rough, and burly, he has been the nemesis of many crooks and lawbreakers in Socorro County. He lives in Magdalena. He has been over nearly every section of the southwestern corner of the State of New Mexico, and knows its rugged terrain as well or better than nearly any other man. He is renowned for his lack of fear, and truthfulness. That is why I believe his account of the Adams Diggings far more than any of the others I have heard. Here is his story in his own words:

"Sure, I knowed old Adams. I knowed him before he left Magdalena, and after he came back. Never was a bigger old liar. Why, he'd tell a lie when the truth would fit better. He was used to braggin' and stretchin' the truth. He was quite a drinkin' man too. I knowed him to stay drunk six months out of the year," (maybe this was an exaggeration, but other people have told me practically the same thing), "and then go on a spree and throw a big drunk the rest of the year.

"It was in the early part of August, 1864, when Adams and about seven other men organized a trappin' expedition and started up in the northwestern part of the state to trap beaver. They started early and intended to get their camp set up before cold weather came. They camped on a little stream not far from old Fort San Rafael, which is now Fort Wingate and has been moved a few miles from the old site of Fort San Rafael.

"Now I don't know this for certain, but I believe from events which I will try to explain later, that just about dark, a caravan from California stopped and threw camp with Adams' party. They had stopped at Fort Wingate two days before and had told the commanding officer that they were transporting between sixty and eighty thousand dollars in placer gold from California to some of the Eastern states. I know that they were never seen after the time Adams' party was wiped out by the Indians, so I believe that they camped with Adams' party and met the same fate.

"I know from Adams' personal character that he was not above ambushing such a caravan. I did not know Davidson, but as he was Adams' sidekick, I believe he throwed in with Adams and the two of 'em made plans to hijack the California outfit and steal their gold.

"An encampment like that, in those days, usually got up an hour or two before daylight, in order to make an early start. It is said that Adams and Davidson made an excuse to go and gather some wood, as wood had been scarce the evening before and they had not been able to obtain a sufficient supply. I believe that Adams and Davidson absented themselves from camp, so they could go down country a few miles and find a suitable place for waylaying the California outfit.

"While they were gone, and it must have been just as good, daylight came, because that is the time when Indians usually attack, a big bunch of Apaches attacked the camp. So complete must have been the surprise, that the white men could not have had a very good chance to grab their guns and defend themselves. Every man in that camp was killed, scalped, and their bodies mutilated, and all their provisions, horses, and mules stolen by the Apaches.

"When Adams and Davidson returned to camp, they must have congratulated themselves on the luck that had caused them to absent themselves from camp. Rummaging around among the supplies, Adams must have found the gold the California outfit had been carrying. As proof of this, I later saw a handful of that gold that Adams had saved when he buried the rest and it was a quality entirely foreign to that part of New Mexico and identical with some I had seen from California Diggin's. The pellets were about the size of a pinhead, up to as big as a pinto bean, and I knew that nobody ever found that kind of gold in the parts of New Mexico I have prospected over.

"After burying the gold in what they considered a safe place, the two made their way afoot, supposedly, to Fort San Rafael, where they said they reported the massacre to the authorities in charge and petitioned aid from the commanding officer to go back and help them relocate a mine they had found and to view the remains of the Indian attack.

"I do not believe this last part, because many years later, I happened to be in Evans Saloon in Magdalena, in March 1890, where Adams, who had been drinkin' pretty heavy, related a story of how he had gone to Fort San Rafael, on a certain day (he mentioned the exact date, which I cannot now remember) in August, 1864, and petitioned the commanding officer for aid to return to give decent burial to the massacred party and offer him and Davidson protection while they tried to relocate a rich gold claim.

"There happened to be an old, retired army officer in the saloon who had listened intently to Adams' story. This man was Captain Sanborn, who was considered a heavy drinker. However, he did not appear to be drunk at this particular time, and he answered Adams:

"'Sir, since the latter part of your speech concerns me, and it is most damaging to my character, I now take it upon myself to refute your statements and call you a contemptible, damned liar. I happened to be the commanding officer of Fort San Rafael at the time of which you are talking. I recall the day of which you speak very clearly and to my knowledge you never set foot in that Fort in your life. It could never be said truthfully that Cap Sanborn ever refused aid to anybody within a week's march of my post who needed it.'

"'Who's a damn liar?'" bellowed Adams. 'Yuh better eat them words, Cap, or me an' you are agoin' to tangle right here an' now. Bigod! I don't like army officers anyway, so I might as well wipe up th' floor with one of 'em right now.' Saying which, he started for Sanborn.

"Cap Sanborn ran behind the lunch counter and grabbed a big butcher knife and jumped over the counter. Adams ran out the front door and Sanborn chased him for a couple of blocks shouting that Adams was the dirtiest liar that ever lived. He could not catch Adams, and returned to the saloon, where he again told everybody in hearing distance that Adams had not ever been in San Rafael.

"From the above incident I drew the conclusion that Adams and Davidson never went to Fort San Rafael at all, but passed a considerable distance to the south in an effort to avoid it. They limped into the little town of Reserve, sore-footed and half-starved.

"It was in Reserve that Adams showed a couple of pieces of ore in quartz form that was exceedingly rich, and stated that it was from the mine he had found before the Indians had massacred his party. He made no mention of the California expedition.

"I later saw the same samples Adams had shown in Reserve and recalled that Adams had showed me one of the samples before he left Magdalena in 1864. He had told me then that he had given an Indian some whiskey for the samples and had promised him more if he would show them where he got the samples. If Adams' story he told in Reserve about these samples had been true, there would indeed have been a substantial claim to his having found a rich mine. This is where all such stories sprang from, and those samples were the richest I have ever seen in my life, and must have come from one of the richest mines ever heard of. But to my knowledge, no ore of similar quality has ever been found, and the Indian who gave the samples to Adams must be long since dead and the place he found the samples will probably never be found.

"Adams didn't dare show any of the gold at that time he had stolen and buried. Therefore he and Davidson separated, Adams going to Pima, Arizona to obtain money and supplies from friends to outfit an expedition to later come back and salvage the gold. Davidson went on a supposed visit to go see some relatives in Louisiana.

"Adams was successful in his attempt to raise an expedition, and he sent for Davidson who returned from Louisiana and the expedition met him in Alma, a little town just south of Reserve. They could not find any gold, and Adams later made several solitary trips in search of it, but never had any luck.

"Several expeditions have been organized and sent forth in an effort to find the Adams Diggings, but all have met with defeat.

"It was in 1918 that I decided to see if I couldn't find the bodies of the men who were massacred in Adams' party. Adams had told me that they had camped about fifteen miles north of three peaks that rose up from the plain and were a considerable distance from any other mountains. I got to thinkin' and the only three peaks I knew of between Gallup and Magdalena were the Tres Montosas, which are only about fifteen miles west of Magdalena. Figuring about fifteen or twenty miles north of there, I went to North Lake. A few miles north of North Lake, I found the bodies of five men, all buried in one hole. I could find no clue to any gold from anything in the vicinity, so I came back to town and reported the finding of the bodies. It is my belief that the bodies I found were the remains of part of Adams' expedition, but of course I can't prove this. But there is one thing I do know. That is that an old fellow I know found about twenty thousand dollars buried about five miles north of North Lake, and only a few miles from the place I discovered those bodies. This man's name is Jose Maria Jaramillo, and this is what he told me. But when I asked him if the twenty thousand was in gold dust, he would not tell me.

"That's the way a lot of these old, 'rich-mine' stories get started," finished old Bob. "I've heard that the definition of a miner is a damn liar without anything but a dang good imagination. You can talk to most of 'em, and dang near ever' one of 'em tells you about some rich prospect they struck. But they're always broke and beggin' a grubstake. If their mines was half as rich as their imaginations, they could take a handpick, and a gold pan and make more money in a month than most bank presidents could by wearin' out a half a dozen fountain pens. It's true that sometimes a prospector does hit it rich, but when he does, he generally don't talk and brag on it, but gets busy and gets some capital interested and starts workin' it. That's my story of the Adams Diggings. It is one of the richest mines in the world in the mind of a danged old liar like I knowed Ed Adams to be, and in the minds of a bunch of old, dream-crazy prospectors who ain't got no more sense than to believe in it."

Lost Treasure

by Manuel Berg

During an interview with Mr. Isaro Pacheco, Ranchos de Albuquerque, he related the following:

"What I am about to tell you happened in the country around Roswell. There were three men, sheepherders, and one of the three was the foreman. They were in charge of a large flock of sheep and they very seldom came to town. At least they didn't ever get to town more than four or five times a year. This which I am telling was told by the foreman, and they all swore that it was the truth. Here is what the foreman said:

"'It was very late one night when I started to look for my two men, Pablo and Carlos. The stars weren't out and a strong wind had come up and I wanted to know whether the sheep were in a good shelter. You see, if they are not sheltered they become kind of wild and we have a lot of trouble gathering them together again. It took me a long time to find the two men because they had also read the signs that a storm was coming up and had herded all the sheep into a little valley.

"'When I did find the men they were trying to make a small fire but were having a little trouble. I helped them start the fire and put some coffee to heat. It was very cold and we weren't through with our coffee when it began to rain and blow real bad. We very quickly stamped the fire out and ran to find some better shelter for ourselves. We had already taken care of the sheep so we didn't bother about them. The three of us must have become separated because I soon heard an awful scream nearby and stopped to try and locate it. I called for Pablo and Carlos but only Pablo answered me and then came running over to see me. Carlos had gotten lost. We couldn't leave Carlos alone because one never knows what can happen to a person alone in the desert so we went back to where the fire had been and began to search. We searched very close to the ground, almost crawling on our hands and knees and even that way I almost fell into this hole. I was very scared when I found the hole but I called out if Carlos was there and he answered my call. Pablo came up to me and then we called to him again to find if there was anything we could do to help him. He, Carlos, said for us to tie a candle and some matches to a cord and lower it and at the same time to light a match so he could see where the entrance was. We did this and pretty soon we felt a jerk of the rope and knew he had found the candle.

"'Pretty soon a dim light came from this hole in the ground but we couldn't see anything from the top—I mean we couldn't make out anything that was down there. I called to Carlos that we would throw the end of a lariat down and then pull him up if he couldn't come up any other way. He said to wait a little bit because he had seen something which he wanted to investigate closer. Next thing he calls up to us and says that he has found a great number of sacks and boxes filled with money both gold and silver. While we were talking, Carlos at the bottom of this hole and Pablo and I at the top, Carlos' light went out and we heard a strange voice say, 'Don't be afraid. Many men have died for this treasure and many more are going to die'—and the voice began to laugh, a laugh so horrible that we almost fainted; that is, Pablo and I, maybe Carlos did faint. Very much shaken I called to Carlos if he had the end of the lariat so we could pull him up. We didn't get an answer right away. Then we saw a light again and Carlos' voice said that he was coming up right away but first he would try to bring one of the sacks of money up with him. This time Carlos tied the lariat around his body and walked further into the cave. Then Carlos said that even one sack was so heavy that he couldn't lift it up. As he said this the candle went out again and the strange voice began to laugh, a laugh that makes my blood run cold and many times I wake up in the night and my heart is filled with great fear. After several moments the laughter died down and the voice said in a deep rumbling tone, 'You will never be able to take one sack. You must take the whole treasure at once or else nothing.'

"'I know that Carlos fainted this time because we felt a strong pull at the lariat and frightened as we were we began to pull it up. It was a very hard job to pull Carlos up because he weighed almost two hundred pounds, and first we had to drag his body across the floor of the cave. By the time we got Carlos out he had come to life again but he was a very sick man. He had a high fever and we got our horses and tied him onto his and came into town. Carlos never did get over his sickness. Within two weeks he was dead. Now Carlos being gone did not upset me a great deal because I thought it might have been a natural death but when Pablo became sick right after Carlos was buried, I became worried. I quit my job and left that part of the country but later I heard that Pablo had also died and his last words were that he had been cursed by an evil voice.'

"Now," Mr. Pacheco said, "that is a story I heard from this man who lives in Martines town in Albuquerque. This man also says that he has a map showing just where this treasure is hidden but I have never seen this map. When I asked him why he didn't go again to get the treasure, he said that since Pablo and Carlos were dead and he still lived, he preferred to go on living and let somebody else find the cursed treasure."

Lorin W. Brown
Santa Fe, N.M.

Date: Apr. 8, 1940
Words: 111

CHILILI

"It was at Chilili that one hears for the first time the stories of buried treasures and treasure hunting expeditions. Less than two decades ago, so the visitor is seriously informed, a Brazilian succeeded in lifting the hidden gold underneath the altar of the old church. For weeks he had employed men digging within the ruins of the sanctuary. Then came the hour when he announced that the following day he would unearth the treasure and would divided it with the people.

During that night, however, his men made away with it. The people of Chilili sought to wreak vengeance on the Brazilian, who remained, but the proposed lynching was averted."

FROM - Twitchell's Leading Facts of New Mexican History.

Gambusinos

by Lorin W. Brown

"Gambusinos" is the name given to placer miners who roam the hills seeking scattered pockets or deposits of gold, mostly working over country which has already been worked on a large scale, trying to retrieve what others have left or overlooked. For the most part, their efforts are poorly paid, but they are a contented lot, and there is always the possibility of a large strike to lure them on.

Near Santa Fe, at Cerrillos in fact, which is near the old Ortiz grant, the greater part of the population are "Gambusinos." At one time a thriving, bustling city because of the coal mines at Madrid, and because Cerrillos itself was the natural center for the other activities which surrounded it, the people did not lack for employment. At one time there was a smelter at Cerrillos, the Cash Entry mine and the others, north of Cerrillos, the big mill at Waldo, the mines at San Pedro and not the least of these which contributed to the prosperity of this city, was the gold diggings on the Ortiz grant. The best known was "La Real de Dolores" from which mine the little mining settlement on the Ortiz grant took its name.

So that with all of the sources of wealth surrounding it, added to the fact that it was on the railroad, Cerrillos was known far and wide for its gayety, its gaming tables and its night life—a typical mining town.

But now Cerrillos falls into the classification, more than any other, of those towns which are scattered all over the Southwest and known as "ghost towns."

There is very little livestock raising in the vicinity and no agriculture to speak of. The mines of Madrid still contribute to some activity at Cerrillos, but the greater part of its one time numerous population have left, and the few who remain live surrounded by the false fronts and gaping windows of its departed glory.

The once productive mines on the Ortiz grants have been closed down because of litigation. In seeking for a means of subsistence the remaining men of the town, remembering the large quantities of gold which formerly had been taken out of the grant, became "Gambusinos" and with rude, homemade dry washers would filter into the grant, on the sly, and wash the dirt which they scraped out of the gullies and washes. In stating that they filtered into the grant, I will explain that the company which laid claim to the grant had prohibited entrance to the grant on the sly, making of them a close mouthed and secretive clan.

The absence of the men of the town was especially noted after a rain, for then they knew that the gold was washed down from the hillsides and slopes into the gullies and arroyos and collecting in the pockets or depressions gave better promise of profitable returns for their efforts.

"Gambusinos" usually worked in groups of three or more for one of their number must be posted to watch for the approach of the care-taker of the grant. Once in a while a rich strike would be made. Two years ago one of these men found a nugget which brought him a little over three hundred dollars. Another one of my acquaintances amongst them located a rich deposit quite out of the bounds where gold was supposed to be. He very quietly moved in with his family to help him, informing his inquisitive neighbors that the whole family was going piñon picking. He succeeded in working the deposit and taking out close to three thousand dollars worth of dust, before he was discovered by a group of his skeptical neighbors and, since the grant was forbidden territory, he could not keep his discoverers from sharing in the rest of the gold left in his claim. I have reason to believe his story, because as I knew then, and as his neighbors attested, after his return he bought a car and furnished his house with new store-bought furniture to replace the homemade benches and tables which had served him previous to his fortunate discovery.

There is one old timer, who has lived all his life in that section and who has been actively connected with all of the mining ups and downs of the grant, who is the only one allowed free ingress and egress to the grant at all times. The reason for this is that he is possessed with the knowledge of the location of a rich vein of gold bearing ore. The owners of the grant hope that someday he will lead them to the location of this vein. But he is a very crafty and wise customer and when conscious of spying eyes he will confine himself to washing very poor paying dirt out of the arroyos or the bed of the little stream, which is found close to Dolores. He makes his camp within the ruined walls of what used to be the stone mansion of one of the owners of the mine, when it was a thriving camp. Here he would grind the ore, which he had secretly brought in the previous night, in a large, old fashioned mortar with a pestle shiny from much use. The resulting powdered rock he would carry down to the stream where he would wash out the gold in his pan. He is still making a very good living from his hidden gold vein and as yet the company has not been able to find out its location. This man it was who conceived the idea of washing the adobes from the abandoned houses in Dolores and securing from them much gold. I would name him the king of the "Gambusinos" because he enjoys more privileges than the rest and has made a much better living from the hills of the Ortiz grant than the rest.

Secrets of the Guadalupes

by Lorin Brown

The Guadalupes (mountains) stretch for 75 miles across the southeastern corner of New Mexico and extend a few miles across the line into Texas. They are thirty miles in width. These figures do not include the foothills, but are rather the range proper.

Their inaccessibility and precipitous canyons have made them a sanctuary for those seeking refuge from pursuit, first the Indian from the Indian, later the Indian from the white man and vice versa. Then those who fled from "Johnny Law" sought its narrow canyons where behind a rock barricade one man could hold off an army.

Yet the lure, which has attracted men for two generations to penetrate its fastness is the lure of gold; buried treasure, bandit loot and Spanish mines.

By far the strangest tale of lost treasure is the one of "Old Sublette's" mysterious source of wealth. Old Sublette, a water witcher for the T. & P. Railroad, was employed by it to locate wells along the right of way and contiguous country, over which the railroad was to be built. This he did with the aid of a mystic willow wand which very accommodatingly would turn in his hand to indicate a subterranean water supply. But his wand must have pointed the way to greater hidden treasures for one day old Sublette came into camp with a sack full of gold bullion. He never lacked for money after that; he would disappear into the Guadalupes, and return with gold whenever he wished. Efforts to trail him were ineffectual, he was too good a shot and too canny to leave any trail.

When he died his secret died with him, although he left a map, revealing its location, to the only friend who was with him when he died. Because of some queer code of honor, this friend never made use of this map, rather he roamed on trails on quests of his own and the secret died with him also.

Nevertheless the certainty and the knowledge of many who had seen Sublette's gold, sent an unending stream of seekers of all descriptions into the Guadalupes' many canyons. Two such, at the end of a discouraging day, were told by a Mexican sheepherder that they were two years late in their quest; that a fellow by the name of Long had found the gold. Express office records showed that a certain Ed Long had consigned $30,000 worth of gold bullion to himself in some city in California.

Later searchers found a shattered redwood box like the ones in which bullion was carried on the coaches of the Butterfield stage line. The Guadalupes are full of caves and caverns of smoke-blackened walls from the smoke of the fires of pre-historic man—mummies, skeletons, and pictographs are other relics of these ancient people found in these mountains; pictographs of the modern Indians, of whose history science has cognizance, are found beside or superimposed on pictographs of a people of whom nothing is known, pictographs similar to the ideograph writing of Oriental races.

One of the narrow canyons has a wall of boulders built across it. The harassed individual who built this refuge wrote the name William Bonney on one of the canyon walls. He was otherwise known as "Billy the Kid."

Both Geronimo and Victorio fled to the Guadalupe mountains with their bands when hard pressed by the U.S. Cavalry; they knew that they would find safety from pursuit in the maze of canyons and many caves.

Source of Information: Excerpts from *Secrets of the Guadalupes* by Carl B. Livingston, *New Mexico H'way Journal*, January, February & March of 1929.

The Lost Juan Mondragon Mine

by Lorin W. Brown

"If my father had known gold when he saw it he would have been a rich man, at least that is what he used to say after he came back from working in the placer mines in the Moreno valley." Donaciano Romero accepted a cigarette and continued, "But when these Americanos were taking out gold from this hidden mine my father had many sheep and there had been no necessity for his leaving home until long after this had happened. So he had no knowledge about gold at the time.

"He summered his sheep in the Canada de los Alamitos on the north slope of the Jicarita Peak. My father told of seeing those men ride up this canyon toward the Jicarita. They always came out at sunrise and, returning, entered the dense forest at sunset or after dark had set in. They certainly wanted to be sure no one followed them I guess.

"At first they had a spotted burro only, which they drove before them loaded with supplies. On one of their trips out, one of the Americanos offered to buy a spotted pony from my father, a very good little saddle horse. From a buck-skin 'taloga,' or pouch full of gold nuggets, my father was offered a large nugget for his pony. But always ignorance is a great drawback. Not knowing anything about gold my father refused and insisted on being paid $100.00 in coined money. 'Then you will have to go with us to Mora for your money, but you are a big fool not to take the gold. It will bring you more than you are asking,' said the Americano who wanted the horse.

"And that's how it was all right. At Mora the gold was changed to money and, just to show my father, the nugget was weighed separately and it was worth $146.00. The Americanos joked my father about this and bought him a gallon of whiskey so he wouldn't feel so badly about it. After arranging for the care of their horse and burro, the Americanos left for Las Vegas and my father, borrowing a burro from his compadre Benigno, returned to the sheep camp.

"It was the custom for those two Americanos not to return to the mine until they had spent all their money. They would visit several towns, Taos, Las Vegas, Raton, etc., on these spending sprees so that many men were anxious to find out where they were getting so much gold. Many efforts were made to get them to talk and to follow them, but they were very careful and nobody succeeded. These men didn't talk when drinking and kept to themselves when sober.

"As for my father and his herder, they never thought of following them because these men went well armed, and besides, my father was well content with his life as it was. Finally at Raton one of these men killed the other during a drunken argument and disappeared to avoid the authorities. He never returned and the mine has never been found. Padre Guerin of Mora spent much time and money trying to locate this mine, but in vain. It must have been a rich placer because father said the gold resembled the placer gold of the Moreno valley. After realizing in later years how rich the mine must have been, he wished God had granted him courage to have followed those men some night. He might have been lucky to have escaped alive and we might have been rich today. Quien Sabe, one does not get more than he is supposed to in this world.

"Look at Juan Mondragon; he fell into the richest mine yet, one might say, died a poor man and denied himself the riches which God wished to give him. He was herding sheep for the Garcias from Santa Fe on the Eastern slope of the Truchas Peaks. One of his milk goats fell into a hole and couldn't get out. Going to her rescue, Juan finished caving in the dirt and pole covering of an old mine. After getting the goat, Mondragon investigated and next time he came home he brought back some very rich ore containing much gold. He also brought some very crude tools, homemade and very rusty. The Garcias, his patrons, offered him half of their three thousand head of sheep and two horses if he would show them where the mine was. But he was an ignorant one and refused. I think he was afraid he would be killed after he showed the mine."

"That man was a 'tonto' (fool)," broke in Donaciano's wife, Eulalia, who had been listening all the while as she patted out tortillas for the evening meal.

"Don't you know he was always fighting with his wife, when he was home. He never let her go to the river for water and he wouldn't take her to the dances. With the company's permission he was a 'pendejo.' Who was there to watch his wife while he was gone? That is the way all these jealous ones are. And didn't he try to hang himself three different times? Twice his poor sister Tonita came running for my father to cut him down, and once, my father not being home, Tonita and I held him up by the knees, while his wife, standing on a bench, cut the rope. That time he almost went. The raw-hide rope was pretty tough, the knife was dull and I think the wife didn't use all her strength. I don't blame her either. He sure had a sore throat for a long time after that."

"Si, a very foolish man," I said. "He didn't have to show anybody this mine. Why didn't he stake out his claim and work it himself?"

"In those times there was nothing known about that," was Donaciano's explanation. "Time after time different parties came to see Juan offering to pay him, to make him rich, promising him anything and everything if he would disclose the location of his mine, but his fear of being killed kept him from doing so. Strange that a man who tried to kill himself should be so afraid of death at the hands of another. No?

"For many summers the slopes back of the Truchas Peaks were full of strangers looking through every valley and corner of the hills for Juan Mondragon's mine, but I guess no one has found it up to now.

"This Juan belonged to 'Los Hermanos de la Luz' or 'penitentes.' At one time due to persuasion of the 'difunto' (deceased) Hijinie Torrez, Juan agreed to show the brotherhood the location of the mine. The mine was to be worked by all, share and share alike and a certain percentage to go to the Morada fund. Only Hijinie Torrez could have persuaded Mondragon to agree to this arrangement. 'Tenia una labia muy suave.' (He was a smooth talker.)

"After a velorio or wake at the Morada the party was to set out. This velorio was in honor of Santa Inez del Campo, the patron saint of those who live out in the open, also the saint to be prayed to for recovery of lost stock, lost persons or anything lost out in the hills or plains. So after the night of the wake these men set out for the hills. Now you will see what happened. It is useless to go against God's will. The first night, while camping at the Brazos of the Rio Medio, four of the younger fellows were talking apart. One of them said, half jokingly, 'As soon as we know where the mine is let's start a fight over the amount to be given to the Morada. Then we can get rid of these old men and the mine is ours. We can pretend some Americanos killed them. Who can deny our story when we get back?' He was cautioned by another not to speak so loudly, but too late, for Mondragon, always suspicious, had overheard and next morning refused to go any further. The young men then, really angry, threatened to hang Mondragon from a tree until he would consent to lead them on. But not even those threats would make him change his mind. The older men advised returning and waiting till later while they tried other inducements on Mondragon. The truth is that they were a little afraid of the younger fellows. Gold brings many bad consequences even before it is in one's hands.

"Later Mondragon used to go secretly and bring back gold and very rich ore, taking with him his nephew, Filogonio, who was then about twelve years old. This nephew always said his uncle did not take him clear to the mine, but that he left him at a certain spot while he went on, returning later with his gold. But I think Filogonio knew very well where it was, but like all the Mondragons, he was very stubborn and would never disclose anything. At different times he would be gone for two or three days and here three years ago he bought all that land across the river. Also there was that land he bought in Truchas.

"You know, in Filogonio's last sickness last year how well that Spaniard took care of him? Well, after Filogonio died the Spaniard admitted that Filogonio had promised to show him the location of the mine in case he recovered, but he was quite sick then. The Spaniard also said that Filogonio had given him signs to go by and that they will be found this summer. Maybe so, but Quien Sabe?

"And what a time Filogonio took to die! Such a long time he was sick. You know a penitente will not lie down in bed when he is sick. He must be kept sitting up. Also they believe that no matter how sick they are they will not die until their feet touch the ground. Filogonio's sister-in-law from Truchas, seeing him so sick said, 'Why do you keep that poor man suffering so? Can't you see he is marked for death? Let his feet down on the ground.' So she helped the wife turn him sideways in bed letting his feet touch the earth floor and that was the last of Filigonio and maybe of Juan Mondragon's mine. 'Sea por Dios.'"

Buried Treasure

by Genevieve Chapin

Traditional, almost to the point of being folklore, are the tales of buried treasure in the Southwest—some small part of it being supposedly in this section of the state.

The Indians believed that gold was just fragments that were left over of the material which the gods used in making the sun—hence, to them, it was sacred, and the little they had was kept solely for their ceremonial rites.

When the Spanish came in, and began digging out this sacred metal, storing it in the ground till such time in the future as they saw fit to dig it up again and take it to Spain, the Indians resented it very deeply. Finally they staged an uprising, killing many of them, and driving the rest back into Old Mexico before they had time to dig up their buried treasure.

Later, at the time of the re-conquest of the country by the Spanish, the caciques, or high priests of the Indians, were killed. This was a double tragedy, for these same high priests had, after the Spanish had been cleared from the country, put enchantments on all the mines and caches of treasure, so that none but themselves could get it.

Thus, with the death of the caciques, perished the ability to secure the gold and treasures, and nothing but disappointment and disaster awaits those who seek to break through this spell.

Many are the tales told of such expeditions. The belief is yet strong throughout the state that such treasure is still buried here.

One such expedition occurred as follows:

A fellow named White came into a camp near Amarillo, Texas one day with a sack of what appeared to be cement, mixed plentifully with gold dust. It assayed about 62-½ pounds to the ton, and the camp went wild over it. He refused so persistently to tell where he got it that they threatened to hang him, so he at last promised to lead them to the spot, and set out with a crowd of them in attendance.

He volunteered to watch the horses that night, so they could get a good night's sleep before setting out, bright and early next morning. When he did not show up at breakfast, they supposed he was downstream watering the horses; but when they were all ready to start, and he was still absent, they instituted a search, but never saw him again.

Realizing he had tricked them, they set out alone to find the mine, but with no success, so finally returned to camp.

Some three years later, White showed up at Salt Lake City with a fabulous amount of gold, but disappeared again after awhile, taking his secret with him.

The natives of the State believe that it is useless to search for any of this buried treasure, as only occasionally has it ever been known to profit the finder, even after he secures it. So-called witches are sometimes consulted with the belief that they may be able to break the spell of enchantment, but rarely does this avail.

So we of New Mexico go our way, tormented sometimes by the memory of these tales, telling ourselves, as have luckless prospectors for years past—"It's there—if we could just find it—." But we can't.

Sources: *Native Tales of New Mexico*, pp. 184-185. *Enchanted Gold* by Frank O. Applegate, published by J. B. Lippincott Company, 1932, Philadelphia, Penn. "Lost Mines and Buried Gold," article by Eleanor Kay, published in *New Mexico Magazine*, September, 1935.

Lost Treasures of Grand Quivira

by Edith Crawford

Grand Quivira is a buried city at the head of the volcanic eruption. Every reader of old Spanish history has heard of the lost treasures of Grand Quivira. Mention is frequently made of it in the old Spanish chronicles and many persons have sacrificed their lives seeking for its hiding place. According to the old Spanish chronicles, Grand Quivira is situated about 175 miles northeast of El Paso, on the western border of Lincoln County, New Mexico. The outlines of the once large city or Pueblo are still plainly visible.

The walls of the old adobe church are still standing and the outlines of the other buildings are plainly discernible. History and tradition of this old town are both corroborated by the surroundings. The old Spanish history tells us that here was located the seat of government of New Spain, where all the wealth of the country was brought and turned in treasure vaults of church and state. According to the Spanish records, Grand Quivira was a large Indian village at the time of the first Spanish conquest and was wrested from its Indian occupants about the year 1616. In 1680 came the great Pueblo Indian uprising when the Spaniards were driven out of the country as chaff is driven before a wind. De Vargas, the writer of the chronicle says: "We traveled to the southward, fighting all the way. We have 270 burro loads of treasure. We buried it between two little hills, burning brush over the spot." The Spaniards were driven out of the country as far south as the Gulf of Mexico. Where are the two little hills De Vargas speaks about? Attempts to solve this problem have cost many an adventurer his life. All around Grand Quivira can be found holes dug in the ground where some poor deluded mortal has been seeking for the lost treasure of Grand Quivira. Every summer, excursions are made to that place. It is a very pleasant place to camp.

Raines, Lester
S-700

216 words

TECOLOTE

<u>Name</u>: Tecolote (Tay-koh-loh'-tay) from the Aztec <u>tecolotl</u>, the ground owl.

<u>Population</u>: 300

<u>Location</u>: On U. S. Highway 85, 11 miles southwest of Las Vegas, on the Rio Tecelote.

<u>History, Development, Points of Interest:</u> First settled by Salvado Montoya, who petitioned for a grant of land there in 1824. It was once the headquarters of Moore and Mitchell, one of a chain of posts established to furnish forage and corn to the army units. The sutlers at Tecelote were supply headquarters for Fort Stanton. The ruins of their headquarters and the stables still remain.

The church was built in 1905.

Two miles below Tecelote, about a mile off the highway, are ruins of two Plains Indian pueblos, owned by the Las Vegas Historical Society. Finds resulting from excavations here are exhibited at the Veeder Museum, Las Vegas.

Inhabitants relate that about 10 years ago a small band of Indians came to Tecelote in search of a treasure buried there by the great-grandfather of one of them. He said that the place had then been called Caco Canon. The Indians camped near the river, built a huge fire before which they danced, and next day disappeared. Whether they were successful in their search for treasure is not known, although the places where they dug were found.

SOURCE OF INFORMATION

Information supplied by N. H. Stapp and Margaret M. Lucero.

Prospector 40 Years

by Edith L. Crawford

The *El Paso Daily News* obtained the following interview from Jerry Hockradle as he passed through El Paso en route to California.

Jerry Hockradle, one of the oldest residents of New Mexico passed through this city today en route from his home at White Oaks, in Lincoln County, to San Diego. Though about 70 odd years of age, he is hale and hearty and a typical prospector in appearance. He said to a news reporter today: "Yes, guess I am the oldest resident of New Mexico, for I went there in '68 when there was nothing in that section but mountains, wild animals and Indians, and there were plenty of the latter. I don't suppose it will do me any credit to tell just how many red fellows I have dropped in my career. In fact I never kept track of them; but I will say this, that after we once got the Indians to knowing us we did not have nearly so much trouble with them as we had with the bad Whites and Mexicans.

"That was about the time that Pat Garrett first became known in that country, or rather when he first got a job as a cowboy and deputy sheriff. We had just gone through with one war between Mexicans because we had sheltered their white enemies, when the Kid came into prominence as a killer. He seemed to have the sheriff pretty well bluffed and the people were getting pretty tired of him anyway, when they came to me and wanted me to run for sheriff to clean him out. I said that I would not take the job under any circumstances, but told them that there was a likely young fellow in a cow camp who was a deputy sheriff, named Pat Garrett, who could do the work. Then we had the convention and nominated Pat. And I will say that he made the best sheriff Lincoln County ever had.

"Now I am quitting New Mexico for good and am going over into the Muddy river country region on the railroad that Senator Clark is building from Salt Lake City to Los Angeles. I passed over this country with a Mormon wagon train, away back in '66 when I was coming east to New Mexico. The place is off the Mohave Desert and when I passed over there, there was alkali so thick on the ground that you could have skated on the crust. Then the water holes were 65, 72, and 85 miles apart as you came to them in crossing the desert to Muddy river, and it was only at certain seasons when you could cross at all or get water, because the wild beasts would drink it before you would reach there. It was on this spot that we had a bloody battle with 150 Indians, Paiutes. The Indians had stolen some of the horses that belonged to the Mormons I was traveling with; when the men of the party tried to recover them they took position on a rocky knoll and fired several volleys into the camp. Our party then made a detour and got behind the rock. After the smoke cleared away we found 19 redskins whom we buried the next morning. It was a narrow escape for us for there were only about six good guns among us. While we were passing over that section I saw some mineral stains and now that the railroad is being built through there it will be a fine place for prospecting. I expect to stop there on my way to San Diego and see what the stains amount to. Then when I reach San Diego I will get an outfit together and go to the peninsula of Lower California where I understand there are some promising placers. I made a good stake out of the White Oaks claims and I hope will have good luck in that country. I know it is there for a cousin of mine told me that he got enough gold out of a flour sack full of dirt to buy several sacks of flour."

Source of Information: *White Oaks Eagle,* dated October 23, 1902.

The Lost Gold Mine

by Kenneth Fordyce

Among the highest and most precipitous peaks of the Guadalupe Mountains in Eastern New Mexico at a spot now unknown to man lies the Lost Gold Mine, known to some as the Lost Sublett Mine. Some of these peaks rise five thousand feet above the plains and as high as nine thousand feet above sea level.

Travel among them is very difficult for even the most experienced. According to tradition the Sublett Mine was the richest in the world.

A Missourian from St. Louis, Ben Sublett, while yet a young man, brought his wife and three children, came across the plains of Texas, and finally got a job in West Texas as a ranch-hand. He remained on this job only long enough to save money and buy a team, a wagon, and an outfit, then he was off to the Guadalupes in Eastern New Mexico, surrendering to that desire which the lure of gold creates. No one knows just how long Ben Sublett searched, nor where all he went, nor where it was he found the gold, but he was successful.

Ben Sublett's lavish expenditures bore out his statements that he had found the richest mine in the world when he returned to the little western town. He had plenty of gold. GOLD! The little town was wild with excitement. Many started at once to hunt for the mountain where Ben had found the precious ore.

Ben Sublett would tell about the mine but never would he divulge where it was located. When he would use up his supply of gold, he would quietly slip off, being sure that he was unobserved, and apparently again visit the mine, for he would eventually return with plenty of gold nuggets. He would then spend freely until that supply was gone before returning to the mine. He provided lavishly for his family and housed them in a beautiful and comfortable home which he built especially for them. On his trips to his mine, he would never take more gold out than he could conveniently carry in his wagon.

Ben's friends pleaded with him to reveal the location of the mine but he always refused. They tried to get him to let some of them accompany him to the mine. Finally he did take his young son, Ross Sublett, and Ross's boy friend, Mike Wilson, with him on a trip to the mine. The boys were small and after that one visit to the mine they had no opportunity to go to the mine until after the father's death.

Ross felt sure that he could find the place; he remembered that it was near a spring, and he said that his father had told him that it was in the Russell Hills of the Guadalupes. He failed to find the place where his father got the gold out of the hole so near the surface that it seemed that he was getting it off the top of the ground. Ross had asked his father during his last illness to tell him where the mine was located but the old man shook his head, smiled, and said that it was no use to try, he could not find it.

Mike Wilson, the other boy, claimed that he found the mine some years later. He produced a large sack of gold nuggets to prove it. He went on a spree to celebrate and spent all of his money, then attempted to return to the mine to get more gold but on his return failed to find the rich spot.

Lucius Arthur claimed success in finding the mine also. Two Mexicans who were supposed to know where the mine was located unknowingly led the way to it, he claimed. He followed them for days and they led him to the spot. He showed up with a large supply of gold to prove his claims. He said that he had broken the gold off of a ledge near the brink of a chasm, the walls of which were so steep that he could not descend. The walls went down some seventy-five feet or more, according to Arthur. He secured ropes long enough to lower himself into the chasm and went back to the spot. He was never heard of again.

The location of the mine remains a mystery to this day.

Source of Information: Re-written from an old newspaper clipping without name or date; Mrs. A. E. McCready, a relative of the Sublett heirs, has the clipping.

Treasure

by Ramitos Montoya

Time and time again I have heard stories about hidden treasures in New Mexico. Years ago when New Mexico was still a plain with wild Indians running all over it, there was much gold in this region. The Indians had possession of this gold. Many of them practiced witchcraft; by some magical power they enchanted much of the gold. A treasure that is enchanted is not easy to discover. In fact, many say that only those practicing witchcraft are able to get the treasure.

The following stories were told me by my old uncle.

There was a certain house at Roy, a small village, where many strange things happened. One night while the family enjoyed their supper they heard the galloping of a horse. The sound came nearer and nearer. The head of the house went outside to inform himself about the rider. Instead of seeing anyone he heard only a voice crying for help.

At another time, laughing and loud talking, murmuring, the crying of women, and singing were heard. The people never knew the reason for such strange happenings.

One night, when the family was returning from a *baile*, they were very much surprised to see a light in the house. As they looked through the windows, they saw a strange sight. In the middle of the room was a table full of lighted candles. As they entered, the table disappeared.

On still another occasion the family was sound asleep. They were abruptly awakened by the clanking of heavy chains. They offered prayers and had the priest bless the house. The house, however, seemed to be bewitched. It is claimed that in this house there was a hidden treasure enchanted by the Indians, and that for this reason strange things happened. The people searched for the enchanted treasure but were never able to find it.

There is another story that I recall about the treasures. The Spaniards had a treasure, which they piled in bags in an old wagon and made their way toward the mountains at Fort Union. The heavy load was pulled by mules. The men were ambushed, killed, and the gold stolen. The five robbers agreed to hide the treasure and take only part of the gold at a time. The treasure was hidden together with the bodies of the murdered men. In order to find exactly the spot, the bandits inscribed some words on a rock: "Dig three feet to the right. Keep on digging."

As time passed, four of the men died. The remaining member of the band wrote a note about the treasure, a note found later by his son. It read as follows:

"There is a hidden treasure on top of the big pine mountain. Dig three feet to the right. At first you will find some human bones but do not be alarmed. I possess part of this treasure. The others that had part were Juan Rubio, Carlos Muno, José Aldano, and Juan Luna. Please tell sons of my above friends about this treasure.

    Your father,
    Bialkin Perez"

The Perez boys, together with the sons of the other men, went to look for the treasure. To keep their courage up they drank whiskey to their hearts' content. They worked all day but did not accomplish much, for they were under the influence of liquor. They soon fell asleep. This time others who were watching what the lads were doing came and carried the treasure away.

An Expedition

from New Mexico Federal Writers' Project

News Item

There is being prepared an expedition to start from Abiquiu on the first of the next month for the Navajo Country to search for some gold mines which is said to exist somewhere in that region. The Indians have for many years reported the existence of exceedingly rich placers in their country but have never let their exact locality be known. For retaining this information among themselves they say they have good and sufficient reasons and no inducement will make them divulge it. The proposed expedition will be composed of between one and two hundred men and will be provisioned and equipped for two months. It will be under the direction of experienced mountaineers and if it is possible to find the placers they will be found by these prospectors.

(Note) There are Navajos around the Reservation today—1941—who are perfectly willing to guide most any white person to a gold mine that their father had told them about. These searches have always proved fruitless but the Indian has had a nice trip, been well fed and possibly received some money for his trouble.

Source of Information: Retold from the *Santa Fe Weekly Gazette*, February 22, 1868.

Buried Treasure

Raines, Lester   August 3, 1936 cl 190 words

THE HAUNTED HOUSE *

For many years the Box S was a trading post. Large sums of money passed through the hands of the trader. One day he received a large amount of money to buy wool from the Indians. The belief was that he had a secreat hiding place for his money. The Indians took their wool and meat to trade. They found the building closed. They waited most of the day for the trader to open the store. At last one Indian woman, called "Red Eye" by the people of Ramah, tried the door. It yielded immediately to her touch and she went in. To her horrors she saw the trader lying on the floor dead. She notified the people of Ramah. No trace of the money or the murderer was ever found.

The house is abandoned now but gold seekers still search the walls for a secret hole, and search under the floors for the hiding place of the gold. The story that the house is haunted by the murdered man, looking for the money, is told by the children of the neighborhood.

* Told by Mr. Tom Fallon. Morgan, Elizabeth, "Brief Sketches of Regional Tales of Western New Mexico," A.M. Thesis, New Mexico Normal University, 1935.

The Lost Sublett Mine

by Katherine Ragsdale

Many years ago a man named William Colum Sublett (Old Ben) came west from Missouri to prospect in the Rocky Mountains. With him he brought his wife, and two daughters. Here in the Rocky Mountains he was unsuccessful so he decided to go to the Guadalupe Mountains and prospect.

"Old Ben" as he was called, took his children to a place in Texas, and while he was prospecting he would leave his children with anyone that would take them.

Time and time again he would come back unsuccessful—then one day he rode into town with a pouch filled with large nuggets. He would remain in town until he ran out of money and then he would go back to the mountains (always alone), stay only a few days and return with plenty of gold.

It was in 1892 "Old Ben" died. He told a man (to whom he disclosed his secret) to tell his son where to find the gold—but Ross was never told.

This gold that was brought in by "Old Ben" was of a peculiar reddish tint found only in gold from California, therefore it is believed that he (Sublett) found caches taken from the stage coaches and hidden in the mountains by raiding Indians. (Through these mountains by Guadalupe Point was—because of the Indians—the most dangerous place on the trail from Tipton, Missouri to San Francisco, California).

Not many years ago a stranger went to Pine Springs and asked if he might camp down near the spring, and Mr. Glover gave him permission to camp there. During the night or early morning the stranger left, and was never seen again. Mr. Glover thought it strange and started investigating—and found a large old trunk that had been broken into.

He took this trunk to his home and on close inspection found the imprint of the double eagle. It is estimated to have had $30,000 in money and bullion.

This was probably taken from the "Sublett Mine." The trunk is now in a museum in Texas.

Source of Information: Miss Virginia Minter, Mr. R. M. Burnet.

Pioneer: Hidden Treasure
(Cutter, Sierra County)

by Lester Raines

Seventy pack burros laden with silver were toiling from Old Mexico to Santa Fe. The precious metal, the accumulation of taxes and gifts, was to be used for the propagation of the Catholic faith.

Men and animals wearied by their laborious journey over streams and mountains had settled themselves to rest for the night near the Rio Grande about three days' ride from El Paso del Norte, when an Indian rushed up breathlessly and informed them that soldiers were in pursuit.

Quickly the silver was dumped into a nearby stream and covered with earth and brush. The travelers saved their lives, but, so far as is known, the treasure has never been located. According to a descriptive map received by Mr. Ross from a convent in Mexico City, which he translated into English, the spring is supposed to be in the Caballo Mountains.

Hundreds of years ago a band of settlers travelling northward were attacked by Indians at Point of Rocks some 20 miles from the present town of Cutter, New Mexico. One man in the party hurriedly collected their gold and valuables in a large dinner plate, and, slipping from the wagon, hid the treasure on the summit of Point of Rocks.

A little Spanish boy was the only survivor of the expedition. The Indians left him for dead, but after they had gone he cautiously made his way back to relatives. Many years afterward he returned, but was unable to find the place where the treasure lies concealed.

Source of Information: These two tales were told to Erma Sowell by old settlers near Cutter, New Mexico.

Still Buried Treasure by Hermione Manning

by L. Raines

In the most wretched section of a border city, where a dilapidated little shack clings to the bank of the Rio Grande River, there lives an old man. His name is John Hawkes, but he is more generally known as Old John. The name Hawkes fits him well, as his large bony nose gives his face a thin, sharp hawk-like appearance. He has lost one eye entirely and the other squints out upon the world with an expression of restlessness and suspicion. His head is set deep within the rounding cleft of his hunched shoulders; his limbs are gnarled and crooked; and a peculiar physical trait enables him to revolve his head much as does a bird.

He has lived alone by the river for many years in abject poverty and filth. So peculiar are his mannerisms and so gruesome is his appearance that his neighbors shun him. He is not at all sociable, talks very little, and has only once revealed to another a secret which he has guarded for years.

A few years ago a young man, seeing old John's destitute condition, made an effort to aid him. As time passed an understanding grew between the two. Old John learned to trust the younger man and at last told him of an experience which occurred while he was employed as a ranch hand years ago.

When a young man, John Hawkes worked as a ranch hand on one of the large cattle ranches which lay along the Mexican border, in that rough and unsettled area known as the badlands. So inaccessible were parts of this country that outlaw bands made it a place of refuge, recuperating from their skirmishes and hiding their loot in perfect safety.

Between the years 1880 and 1885 the first railway was built into the interior of Mexico. Travel increased and the transportation of goods between Mexico and the United States flourished. Banditry as well throve and train robberies became frequent. In one hold-up the Mexican government lost $500,000 in gold bullion en route. The robbery was neatly done and for many years no trace of bandits or gold was found by officers of either country.

Old John related that one afternoon about this time he and several other ranch hands were quietly working in the corral when five armed men rode in, drew guns on them, and demanded their surrender. The horses of the bandits and the three heavily loaded pack animals with them showed every evidence of a hurried, fatiguing journey.

The ranch men surrendered to the intruders and were tied. Four of the robbers took fresh mounts, got picks and shovels, and rode away into a nearby canyon, leaving one of their number as guard over the prisoners. In a few hours they returned, released the ranch men, and demanded that a meal be prepared. After they had eaten they mounted and rode off quickly.

Realizing that efforts at pursuit would be useless and probably fatal, the ranch men made no attempt to follow the party. Late that night, however, John Hawkes saddled his horse and rode toward the canyon, following the train the bandits had taken in the afternoon. He had barely reached the entrance when he became aware of sounds on the opposite side of the gorge. He stopped his horse abruptly and listened. Clear and unmistakable came the squeak of saddle leather, the rolling of pebbles displaced by movement. John quietly dismounted, left his horse standing with reins dangling, and crept down the trail. Over the rocky uneven ground he went, through the dwarfed trees to the bottom of the cut, and up to the other side, stopping to listen for sound and stepping firmly so that he rolled no pebble or rustled no brush. As he neared the top he took greater precautions, hesitating longer before inching himself along, concealing himself behind the scrubby little bushes along the canyon rim.

At last he reached a point where he could see what was taking place. Digging and prying among a pile of rocks he saw a man; he recognized him as one of the bandits. Finally, after much effort, the robber dragged into view an object resembling a sack. Several more followed. They seemed very heavy, for the man found it difficult to load them on the horse. It took much pushing and hauling to get one sack into the seat of the saddle. Then the horse was led away. John noted the direction but could not tell exactly how far the man had gone. He waited. Soon he heard the man and horse returning. The saddle was empty. Another sack was loaded on the saddle; another trip was made. After a succession of trips all the sacks were moved. From the last trip the man returned, riding the horse and continuing down a trail along the edge of the canyon which avoided the ranch and joined the main trail several miles beyond.

Old John easily found where the second cache had been made. He opened a sack. It contained gold, shining yellow bars several inches long and an inch thick, each one bearing a number which stamped it as government property. He transferred the gold to a new hiding place of his own but did not report his find, fearing the government would not only take his precious treasure but would imprison him as an accomplice to the crime.

Many times since that night years ago he has moved the bars of gold, often dividing the hoard into smaller caches. Only once has he made an effort to use the gold. With great difficulty he melted a part of one bar and disposed of it for a considerable sum of money. The sale, however, did not pass unnoticed. Old John was rigidly questioned by government officials. He gave them no information which would furnish them a clue to the location of the gold. He knows, nevertheless, that the officials suspect him and watch him constantly. Helpless, destitute, he drowses in the afternoon sun before his ramshackle dwelling, dreaming of his buried treasure. In the meantime a local charitable organization assumes the responsibility for his support.

The Church of the Golden Bell

by L. Raines

One evening as a young Spaniard was riding home along the old trail from La Mesa Escrita through the "mal pi" (pais) south toward Acomita he heard the sound of a church bell. He paused in surprise, and there in front of him, dimly outlined against the cloudy southern horizon, was an old mission.

"Ah! I shall be rich," he thought. "At last I have seen the church and heard the bell. I shall find the gold." But, even as he gazed, the darkness blotted the vision from his view. He leaped from his horse and in the fading light carefully marked the spot on the trail. Who can recognize a landmark made of "mal pi" rock in the badland trails? In the daylight he searched in vain for the hastily constructed guidepost.

The gold of the church with the golden bell has been securely hidden for more than three hundred years. Legend tells us that during the years of colonization a small settlement was built around a mission in the midst of the lava beds of what is now Valencia County. In the mission tower was placed a bell. The settlers, fleeing from the community, buried their wealth beside the trail. If one follows the trail from Inscription Rock over the lava flow south to Acomita and watches carefully on the right hand when the trail turns east he may find the cache buried in a crevice of the rock. The trail is there to this day; several claim to have seen the mission; others claim to have heard of a map locating the cache; but the pines have grown old and fallen, destroying landmarks, and the bell is heard only by some lone wanderer or the mission is seen in the cloudy distance by some hurrying traveler.

Source of Information: Told by Thomas Fallon to Elizabeth Morgan. Morgan, Elizabeth, "Brief Sketches of Regional Tales of Western New Mexico," A. M. Thesis, New Mexico Normal University, 1935.

The Story of Adam's Diggings

by L. Raines

This story was given to me by Duane Aul, son of a rancher, who has lived near the scenes of the events which he narrates. He has retold the story as he has heard it many times around the fireside on winter evenings.

Many years before New Mexico became a state, Adam and his partner worked a mine in the Zuni Mountains. As soon as the men had mined a few hundred pounds of gold, they loaded it on donkeys and started south to trade for supplies. The trail that they followed led from San Rafael, a village a few miles west of Grants, New Mexico, to Gold Spring, where they always camped. Covering a distance of several miles, they passed the next night in the home of an old Spanish woman at Tinaja. When they left Tinaja no one knew the destination except that the trail led south to some city perhaps below the border. The trip usually took at least two months.

After one of these southern trips, which he had made alone, Adam failed to return. His partner went in search of him but found no trace of him in the South. When he tried to return to the mine he could not find the trail. He settled in the Tinaja valley, but after about twenty years' meager existence on the income from the ranch he decided to make one more effort to find the mine. Accompanied by a friend, he searched for weeks and at last was rewarded by finding an old trail sign. With the hope of regained wealth once more stirring in his mind he camped at the old sign just five miles from the mine. He told his friend that he was sure that he could lead him to the mine the next morning, for he recalled clearly that the trail led to a crevice in the rocks and down the crevice to a box canyon and in the canyon was the mine. During the night he became ill and died.

Since that time several parties have searched for the mine and the rumor is that there is a well marked map of the trail in existence. One party reported that it found a box canyon into which it descended with ropes over a hundred foot ledge. On the bottom of the canyon were found a copper kettle, a pick, a shovel, and other articles. No trace of gold has ever been found.

Source of Information: Morgan, Elizabeth, "Brief Sketches of Regional Tales of Western New Mexico," A. M. Thesis, New Mexico Normal University, 1935.

The Treasure of Punta de Agua
by Edna Shaw

by Lester Raines

Punta de Agua is a small village in the foothills of the Manzano Mountains. It had been settled by an Indian tribe, who had occupied it until they were forced through a bloody battle to surrender it to the Spaniards.

The Spaniards knew that the tribe had been one of the richest in the country and for generations longed to know where the Indians kept their treasure. They had, however, been unable to find out; and, as years went by, interest in the treasure gradually died out. The Spanish rebuilt the town in Spanish style. Any race other than Spanish was prohibited from living there. Especially did they object to any Indian even entering the town.

For this reason there was a considerable disturbance in Punta de Agua one day when an Indian came dashing into the village one afternoon. He was without fear. His assurance, in fact, was so pronounced that the Spanish inhabitants were undecided as to what to do with him.

The Indian managed to talk with the priest and tell his mission. He told the priest that he had a map showing where the Indians had buried their treasure. If the Spaniards were willing to help him find the treasure he would gladly take only a small portion of it and give the rest to the town for its own use. He wanted part of the treasure, for it had been his grandfather who had collected most of it.

The priest called the town council together to discuss the question. After long deliberation they decided it would be profitable and possibly their only way of getting the treasure, so they finally told the Indian they would help him.

After they had toiled for days in order to reach the treasure the

Indian said that he would finance a baile for the village and see that they had all kinds of drinks. They were to have a good time at his expense. He was moved to this step in appreciation for they had helped him willingly.

The night of the dance came. All the villagers were having a very good time and enjoying their abundance of liquor to the fullest extent.

They were all drunk before the evening was over.

The following morning the men went, bleary-eyed, to resume their digging. To their amazement the Indian did not come that morning to help them. In fact, no one had seen him since the night before. They started to go ahead in their excavation without him, but when they started to dig they found a big hole was there and that apparently a trunk or big box had been removed. The treasure was gone!

The villagers set out in frantic pursuit of the Indian, but as no one knew in which direction he had gone their efforts were without success. They had done all the labor and received no reward. They have not yet forgotten the incident.

Source of Information: N. B. This treasure story is apparently a cherished one in the Manzano Mountains. Frank Applegate retells it in *"Native Tales of New Mexico,"* although in this case the one who recovers the treasure is a Spaniard, not an Indian.

Raines, L.   152 words

Buried Treasure

Near Tinaja there was a large slab of rock with a round hole in it. This slab stood upright on a sloping hillside in such a manner that at certain hours of the day the sun shone through the hole and a spot of light fell on the ground. It was believed that this sunbeam marked the location of a buried treasure. A young Spaniard, determined to uncover this treasure, carefully marked the position of the spot of light and dug. He struck solid rock. Widening the area of his search he continued his excavations and found nothing but a ledge of rock. He threw his lariat around the "hole in the rock" and struck his horses. The animal dashed forward and overturned the slab.

"Hole in the Rock fooled me but it will never fool another," he said.

*Translated from Spanish story as told by Estaban Back Tinaja. Morgan, Elizabeth, "Brief Sketches of Regional Tales of Western New Mexico," A. M. Thesis, New Mexico Normal University, 1956.

Indian Fight in the Floridas

by Betty Reich

In 1882 a band of prospectors made camp one night in the Florida Mountains, which are about fifteen miles southeast of Deming. They were camped about a mile and a half from Bear Spring, also in the Floridas.

Victorio, the famous Apache chief who had been driven out of the State of Chihuahua in Mexico (just south of New Mexico) by Terrazas—the Governor of Chihuahua—was camped at this spring with a band of his Apache warriors. Neither party knew of the presence of the other.

The next morning the prospectors set out for Bear Spring to get a supply of water, for there was no water where they were camped. The spring was in a cove and could not be seen until you reached the opening which led to the spring.

The Indian braves were cooking breakfast when the leader of the prospectors went into the cove. He immediately turned to the rest of the men who were just behind him and shouted to them to run. The Apaches opened fire and killed one man and wounded two others before the prospectors could reach a shelter of rocks. Victorio led the attack against the prospectors. He was riding a white horse.

The fighting kept up most of the day with the prospectors holding their own. In the afternoon two of the prospectors fired together at Victorio and his white horse. The horse fell and the Indians fled. Not long afterwards the prospectors saw the Indians going south carrying a wounded man.

The prospectors buried their companion and placed a pile of rocks over his grave. Then they started for Fort Cummings (35 miles away) with four of the men who had been wounded in the fight with the Indians. They heard later that eight of the Apaches had been killed and ten wounded.

Squaws at the San Carlos Reservation said Victorio was wounded in the fight at Florida and was taken south to Lake Palomas in Mexico to be nursed, but died from his wounds. However there are different stories as to the circumstances of Victorio's death.

Indian Stage Coach Robbery

by Betty Reich

There have been many stories told of hidden gold, but few of them have any foundation in fact.

One story told is that Indians robbed a stage coach on the Butterfield Trail, near Deming, that was carrying passengers, gold bullion and merchandise. It is claimed that the Indians hid the gold in a cave in the mountains. It is said that bolts of calico have been found in the mountains that were taken from the stage coach with gold.

Many people have explored the mountains in the vicinity of the alleged robbery, but the gold has never been found.

Source of Information: Mrs. Alex Thompson, Deming, New Mexico.

The Lost Mine

by Betty Reich

The Lost Tayopa Mine was a mine in the Sierra Madre Mountains of Mexico (there were really four mines). It was worked by the padres, who combined mining with teaching of the Christian faith to the Indians. After spasmodic hostilities, 3000 Indians gathered at two white peaks that commanded a view of the mines and the fort which the padres had built. The padres fled to a nearby cave with 30 mule loads of bullion and the church bell. After running out of food they attempted to escape and, with the exception of one, were murdered.

In 1896 a Mormon who was a veteran prospector in the West was prospecting in the rough country of Chihuahua near the Sonora border (southwest of Columbus, New Mexico). Americans, Mexicans, and renegade Apaches made the country unsafe for prospectors.

While prospecting, he found the four mines which formed the basis for the Tayopa legend. He found the old fort walls and the gate posts with the crumbling boulder bearing the carving "Tayopa."

He made friends with three Spaniards wearing old fashioned dress and carrying old type Spanish arms, who arrived after he did. They showed him the old manuscript written by the young padre who escaped.

He found everything except the 30 mule loads of bullion and the old church bell which the padres were carrying away.

The Schaeffer Diggings

by Betty Reich

During the '70s the Apaches led by their chiefs Cochise and Victorio terrorized the inhabitants of southwestern New Mexico.

In the fall of 1872 a band of wood choppers was sent from Fort Cummings, near Cook's Peak about twenty-five miles north of Deming, to the headwaters of the Mimbres River. About a dozen soldiers were sent with them as a guard.

Jake Schaeffer, a German, was the cook for the outfit. He knew very little of the woods or of hunting.

One day when they had been in camp for about three months, Schaeffer wanted to go out with the scout, Young, to try and get a deer.

Schaeffer did not return from the hunt. He was tracked to a deer that he had killed by Young and another man, but there his trail was lost. The searchers were attacked by the Apaches and Young was killed. The survivor returning to camp found the Indians had fallen upon it and killed all of the woodchoppers and two of the soldiers.

The others decided to return to Fort Cummings. Soldiers were sent out from there to bury the dead and to search for the Indians and Schaeffer, but they were able to locate neither the Indians nor Schaeffer.

Schaeffer wandered for miles through mountain and desert, finally coming to Fort Craig, near the Rio Grande, almost naked and out of his head. He had no shoes nor gun, but he still had his haversack with him which had many nuggets of almost pure gold in it.

Suddenly he started running and was out of sight before anyone knew what he was doing. Soldiers were sent out to look for him and when they found him the rest of his clothes and his haversack were gone.

Schaeffer was ill for weeks and when he recovered he could remember very little. He remembered that he had crossed a desert and had seen a mountain with a picture of a woman on it.

Neither Schaeffer nor anyone else was ever able to locate the place where he got the gold nuggets.

Negro Diggings

A negro trooper found coarse gold in a gulch in the mountains. He located the spot by a mountain with the face of a woman on its side.

When he was discharged from the army he went back to hunt for the gold, but the day after he made a rich strike he was driven out by the Apaches.

All of the numerous legends of lost diggings in this section are similar. They are regarded by most people as "tall tales."

Source of Information: *Black Range Tales* by James A. McKenna and personal interview with the author.

Jessie Martin, Desert Rat

Told by N. Howard Thorp

Gold was found on Las Huertas Grant at an early date. It was placer in character and was taken from an arroyo north of the town, the mouth of which enters the stream at that point.

A bar of gold a few years ago worth more than a thousand dollars was ploughed up in a field in Las Placitas and sold to a merchant at Domingo. The origin of this bar is unknown, but the old timers say it was gold that was washed out of the arroyo above mentioned and later melted into a bar. This bar of gold was probably hidden during the time when Indians were on the warpath, and whoever hid it probably never lived to return and claim it.

North and east of the town, across the creek, some copper and high grade lead have been found, and the vein doubtless warrants further development, as some recent assays are very encouraging. There is also a well-defined vein of mineral of a yellowish-red color, about fifteen feet wide, running nearly north and south through the middle of the town's only street. This is easy to trace.

Most of the people living in Las Placitas derive their living from the little fields they cultivate and a few have cattle and goats. They also derive a small income from hauling firewood to the towns, gathering piñon nuts in the fall, and working at freighting for the stores and mines.

Sometime about 1880, a typical old "desert rat" named Jessie Martin, driving a couple of burros loaded with camp equipment, picks, shovels, etc., arrived one evening at Las Placitas. He was a wonderful talker, and claimed to have spent the last six months in prospecting the surrounding hills and canyons. A hospitable Mexican family invited him to stay the night, which he did.

There is but one street in Las Placitas. Stepping out of the house the following morning, Jessie saw what he said was a rich vein of ore, and proceeded to locate a couple of mining claims that practically covered the entire town.

Great excitement prevailed among the villagers, who at first intended running Jessie away, but as he faithfully promised to give them work and make them rich right away, he soon owned the town. Telling everyone to keep the find a secret, he employed several men to sink a shaft in the main street. After it was down a few feet, he sacked up a lot of samples, and at Bernalillo took the train to Santa Fe. There he appeared before General Lew Wallace, at that time Governor, and showed him his ore and also some rich assays, which ran into the thousands in gold. He also invited the richest jeweler in Santa Fe, and between him and Governor Wallace, Jessie was soon enriched by a roll of bills, and was told not to say anything about the find. But with one hundred dollars or so in his pocket, and many saloons beckoning, Jessie could not keep his mouth shut, and soon hundreds of people were flocking to the new El Dorado. Colonel Francisco Perea, one of the old rich Dons, and a physician were caught in the tide and swept onward with the excitement. Finally, Governor Wallace and the jeweler, dressed like typical old-time miners, appeared upon the scene as Jessie's partners. This only added fuel to the flame, many thinking that anything the Governor endorsed was genuine.

Everything concerning this boom seemed straight; Jessie had been seen to dig up the samples and return with assay reports running into the thousands. The next thing of importance was the arrival of several prosperous looking Easterners, who made a bona fide offer to Governor Wallace and his partner of several thousand dollars for their interest, but they refused to consider sale.

The news of the strike spread rapidly over New Mexico; people appeared from everywhere. Newspaper correspondents began to arrive, and they kept the telegraph wires busy. You could hear stories that all the houses in Las Placitas were built from the mineral vein and were so rich in gold that the people had all moved out, torn down their houses, and shipped them to the smelter. These assay reports that old Jessie had obtained were posted with the location notices on one of the claims. A peculiar thing about the whole excitement was that the deeper the shafts went, the better the ore looked.

Old Jessie strutted around dressed in boots, gaudy britches, a flaming red shirt, and a wide-brimmed hat, the picture of a successful mining promoter. He met all new arrivals, and after giving them the glad hand told wonderful stories of the richness of the mines and what the future held. An enormous smelter was to be erected at once and a railroad built to tap the main line at Bernalillo, which would bring in coal for the furnaces; however, it did not occur to anyone that the coal from the nearest mine at Madrid or Cerrillos was totally unfit for smelting purposes. To a man like old Jessie, who at the time had a hundred dollars or so in his pocket and the prospects of more, a little thing like a half-million-dollar smelter, and the numerous intricate questions involved relative to the constructive of same, were mere details.

Jessie, as I said before, was getting along in years, an old "desert rat" who time had taught much concerning prospecting and minerals, but little about women.

There was a beautiful native girl then living in Las Placitas of some fifteen or sixteen years of age; and after one glimpse of her, Jessie commenced to hurl presents in her direction. The courting seemed difficult, as the young lady in question already had a beau of somewhere near her own age, who was very much in love. A pair of horses and a wagon and various other presents to the prospective father-in-law at last turned the scales in Jessie's favor; and after a suitable time, the day of the wedding was set and everyone in camp was invited to the festivities. Several sheep and cattle were barbecued for the occasion, while vino del pais was furnished by the barrel. A good local orchestra consisting of a violin and a guitar furnished the music for the grand baile, which lasted until sunup the following morning.

But Jessie's happiness was of short duration. He soon found his bride receiving attentions from her former suitor, so Jessie sent his bride and entire family home. They thought it a great hardship to be thus cut off from their source of supplies, as Jessie had supported them all since the wedding.

About this time, Governor Wallace and his partner seemed to lose their interest in mining, so with a goodly supply of samples of ore they left for Santa Fe. After sending samples to Denver, El Paso, and their home office, the returns showed not a trace of gold, and then they commenced to smell a rat.

The inside facts of this famous fiasco have always been a complete mystery. Everyone who has been interested investigated each angle carefully. The assays were genuine. Jessie's principals had not given him a cent after he had turned in his reports and received a hundred dollars, but during the height of the excitement Jessie had spent several thousand dollars. Where did he get it? No one has ever found out. However, Jessie Martin stayed at Las Placitas after all the strangers had left, continued to work his mine, and always had plenty of money. This went on for several years.

There have been several explanations for this mining blow-out. One is that in this large vein there were pockets containing the ore, and these Jessie high-graded. I believe the ore that was sampled and showed such wonderful returns came from another mine and probably was high-graded from another dump of proven value. But whatever the explanation, there is no doubt, as one of the villagers said, "It was one big excitement while it lasted. One fellow got a wife, even if he could not keep her long, and a lot of other fellows got plenty of mining experience."

La Mina Escondida: "The Hidden Mine"

by N. Howard Thorp

The old town of Quemado "Burnt Place" whose Post Office is Cordova, named after one of the old families living there, is located on the Senate del Rosario Grant, in the extreme southeastern part of the Rio Arriba county, New Mexico.

This grant was made by the King of Spain while New Mexico was still a Spanish province.

Six miles north of Chimayo and some three miles south of Truchas lies Quemado, a rough Mesa country in the foothills of the Sangre de Christo range.

These little mountain towns have changed but little in the past hundred or two years. Everyone is related to his neighbor, and a stranger must not make any disparaging remarks concerning anyone in these towns, or you will be almost certain to be treading on the toes of someone's near relative. Old customs and ways are radically different from those seen in larger New Mexican towns, where, the Americans being in the majority, the natives are suffering from copying their ways and manner of living.

The foothills surrounding this little town are covered with cedar and dwarf pinion, while high in the mountains one meets massive groves of pine. The clear streams, which descend from the high country, leaves one in no doubt why the original Spanish settlers chose the Mountain locations for homes in preference to the plains below.

Always assured of water for their fields and livestock, and with their gardens, goats and milk cows, they were assured of a life with contentment and ease.

The little settlements—especially in the old days when the people had no newspapers—greatly enjoyed the time honored indoor occupation of gossip. Little happenings, things which in no way would interest people in touch with the outside world, was to these isolated people, news of the greatest portent. Superstition supplanted news, and tales of ghosts ("Brujas") were extracted with the knitting needles from the old Ladies' sewing bags.

This district—like most northern and western New Mexico—is a home of the Penitentes, "a sincere people" whose Moradas, Crucifixes, and Descansos dot the hills and canyons. In this setting lies the hidden Mine of Quemado, and the gruesome tragedy enacted, concerning its possession, eventual loss and the death of the three figures most closely connected with its disappearance.

These Quemadenos raise much chili and onions, taking them on burros to Taos and other northern towns, and trade it for wheat and grain which they themselves did not raise. Such a trip with burros was an event to be looked forward to and long remembered, especially so when the wife and family were taken, and given a chance to see their primos and other relations. If it happened to be Taos they were visiting on San Geronimo day, the great feast day of the year, the Quemadenos might possibly prolong their visit for a week.

The central figure connected with, and the original discoverer of La Mina Escondida, was a little slender man named Nepomoseno. He owned a small ranch and a flock of goats—a couple of hundred in number—who supplied him with what milk, cheese, and meat he needed. A widower of some years' standing, but living alone, and the neighbors noticed, quite comfortably. Presently it was observed he would take his burros and disappear for a few days, in the meantime getting some neighbor's boy to herd his goats. Returning from his trip, his burros would be loaded with food, while Nepomoseno would have a new pair of pants and on his head a fine sombrero. As he took the burros to town unloaded, with what did he buy all his fine clothes and food?

Neighbors' curiosity was aroused to a high pitch as to how Nepomoseno had suddenly become so prosperous, and many a young girl began to cast languishing eyes at him.

Few in the little settlement could afford canned goods on their table, and store butter and bacon, canned syrup; in fact, to the mothers of marriageable daughters, Nepomoseno had suddenly become a great catch. Now, a man of sixty can't withstand the adoration and attention of every young girl in the settlement, and so after the assault had been carried on for a few months he capitulated, and chose a black eyed girl some twenty years of age named Emilia Martinez.

After the wedding festivities were concluded Nepomoseno and his wife Emilia, each mounted on a burro, started for Santa Fe to visit relatives. While in Santa Fe Nepomoseno confided in an old friend, "an American who still lives there," the source of his wealth. As he told the story, within a few miles of his Quemado ranch he had found gold, a small sack of which he proudly displayed to his American friend whom I shall designate as J. H. Nepomoseno said he would drive his goats out to pasture, and after arriving at his gold mine—which was in the canyon—would fill his buckskin sack with gold, then cover up all trace of his digging and before leaving drive the goats back and forth over where he had been at work, and so destroy all evidence of the whereabouts of his mine.

Every two or three weeks he would go to Santa Fe, always bringing back for his bride presents, new shoes, new shawls, bright ribbons and what other little things she might desire.

By now his neighbors were aware that he either had a rich gold mine, or had discovered a hidden treasure, so one after another began to follow him as he took his goats out to pasture. But old Nepomoseno was aware of being followed, so after he had taken his flock for some distance from the ranch, he would backtrack, and if he found any sign of having been followed he would not then go near his mine.

Time went on, and Nepomoseno being most of the time away from home with his goats, his wife became lonesome, and deciding Nepomoseno was a little shop-worn found herself a lover. This friendship becoming stronger and stronger, they decided that if the old man was only out of the way, how fine it would be to get married. But hay Dios, the boyfriend had no way to support a wife, and the old man was such a good provider, if they only knew where Nepomoseno got his gold? At last the friend determined to trail Nepomoseno, and if he discovered the mine, they would then decide what to do.

Hard weeks of trailing failed to disclose the coveted location of the mine, until late one evening, and far from the ranch—with the goat flock bedded on a side hill—the boy saw Nepomoseno digging in the bed of a canyon. There undoubtedly was the mine, the source of the old man's wealth. Overcome with joy, the boy ran the entire distance back to the ranch. Emilia, he called, I have found the mine, and breathlessly described what he had seen. But, asked the cautious Emilia, can you again go to the place? With my eyes closed, replied her friend.

A few days later, Nepomoseno returned home, and was almost immediately taken ill; all the customary remedios having failed, the Priest was summoned, and after offering what comfort he could, Nepomoseno passed away. Now the young people began to hunt in earnest for the mine. The boy knew he could go straight as an arrow to the place where he had at last seen Nepomoseno digging, and the goats bedded along the sides of the canyon, but hunt as he would, he could never find the place. Not long after the above happenings, the boy got into a fight, killed the man he was quarreling with, and was sentenced to the Penn for life where he eventually died. Years passed, and Emilia lived at the little ranch farming it and tending the goats, but always expecting to again find the mine. Now to add to her lot, she was taken down with inflammatory rheumatism, with which she was confined to her bed until death. Before she passed away, she confided to her family that she had poisoned her husband by soaking the heads of old fashioned sulphur matches in the coffee which she gave him to drink.

After the passing of Emilia in 1889—the third one to have met death—the country around Quemado was systematically searched. Hundreds of people combed the hills and canyons, beckoned on by that finger of gold. There was no question then, as there is none now, but what Nepomoseno, somewhere in those hills near his ranch, had found a rich deposit. Too many people had seen the gold, and accepted it over their counters in payment of goods. Every once in a while, someone will recall Nepomoseno's mine, and another prospector will return, having failed to find La Mina Escondida.

The End

Lost Mine of the Pedernal

by N. Howard Thorp

The story of the lost mine of the Pedernal has been handed down by the Spanish people from generation to generation.

These Pedernal hills are of pre-Cambrian formation, and parts of old ridges that survived far into Permian time. The legend still survives that these Pedernal "flint-rock" hills hold in their embrace the mystery of a lost mine. Situated in TSP. 7. No. range 12 east in the northeast corner of Torrance County, entirely devoid of timber, they present a barren and gruesome appearance.

Some thirty-five years past, while living at Duran, a sheepherder brought me a piece of quartz which he had picked up at the Pedernal hills. Upon being assayed, it proved to run over one hundred dollars to the ton in gold. Sometime after this happened I was called to Santa Fe. As I made a late start and a snowstorm had set in I made slow time, passing through the Pedernal Peaks a little before sundown. The trail which passes the peaks goes between the highest ones, and glancing down my eye was attracted by a piece of fine looking quartz. I dismounted, and picking it up, put it in my coat pocket. Night coming on I decided to quit the trail, so headed west a couple of miles to the ranch of Frank Gomez, a sheepman living in Red Canyon. During the evening we discussed—that which was uppermost in my mind—the sample of gold-rock, which I had found in the trail.

Mr. Gomez—who had lived many years at his ranch—told me that time and again prospectors had come in search of the lost mine.

One of the stories he told me related how a Brother Elsmere from Bernalillo had spent two years driving a tunnel to cross-cut what he supposed was the gold bearing vein of the old mine; though after running the tunnel some two hundred feet, he gave up in despair.

The following morning I started riding through the snow toward

Santa Fe, but the going was so heavy it was late at night when I arrived at the Station of Lamy, some fifteen miles south of Santa Fe; here I left my horse in care of a Mexican friend and took a late train for the Capitol. The following morning I had my sample assayed, and the returns showed one hundred and eighteen dollars in gold.

Two days later I stepped off the train at Lamy, and saddling my horse decided to follow the new railroad, which was then being constructed from Kennedy to Torrance; this was known as the S. F. Central railway, and at the time I speak of, December 1902, it was only partially completed.

Stopping the first night at a ranch-house near Moriarty, the night following found me at Willard, at that time consisting of a Company store, small Station, and bunk house. Being furnished with some bedding, I found my way to the bunk-house, where for company there were already two men, who after we became acquainted told me they were "prospectors" from Colorado.

It puzzled me, wondering what prospectors could expect to find—in the way of mineral—in the Estancia Valley.

Presently one of them asked if I was well acquainted with the country, and replying in the affirmative, the other wished to know if I could direct them, how to go to the Pedernal Mountain. I was now interested in knowing what they were after, and before we turned in, they gave me the following information.

It seemed that Willard S. Hopewell, who at that time was building the railroad, had been shown by a store keeper in one of the Manzano Mountain towns a piece of gold ore, which had been picked up in the Pedernal Mountain by his brother, who had been herding sheep there during the previous summer. Mr. Hopewell had an assay made of it: the specimen ran over a hundred dollars to the ton in gold. Not wishing the find to become known, Mr. Hopewell sent to Colorado, and employed these two prospectors to try and locate the vein.

From the bunk-house door the following morning I pointed out the peaks of the Pedernal Mountains, which from our view-point could plainly be seen.

Bidding my friends goodbye, I headed for Duran, feeling so rich I promised my old horse he would never have to do another day's work.

Arriving late I told my partner of my trip, and how everything pointed to a lucky gold strike.

We had at that time working for us two miners, who had taken a contract to sink for us a well. As from the surface down it had been hard rock, they were not making much progress, and like my partner and myself were pretty well discouraged, and about ready to quit.

When I told them of our prospects, and offered to grub-stake them and work on a fifty-fifty basis, they jumped at the chance, as the single word gold to an old miner's ear is the sweetest word in the English language.

We quickly loaded our buck-board with tents, tools, giant powder fuse and caps, water kegs, and grub for a month. Hitching our little mules Sunspot and Brigham Young to our out-fit, away we went. Feeling the way we did it would have taken a very sizable bank-roll to have bought out our prospects of fortune. After leaving, we were somewhat worried that the two prospectors I had left at Willard would beat us to our discovery, but I silenced my partner's fears by mentioning that the Pedernal had been prospected for many years, and although it was known from the legend that at one time a rich mine existed there, none of the later days prospectors had been able to locate it.

The two old prospectors, Ed Davis, and Barney Carney, were well along in years, and had spent most of their lives in going to get rich, and never had, and like all who have been disappointed, are inclined to be skeptical.

After arriving at the Pedernal, I took them through—where I thought I had—picked up the specimen of gold, and although the whole pass through which I had ridden was covered with quartz, there was no particular piece of rock I could find which looked like the one I had assayed. However, we made camp, cooked dinner and—as I had brought along a saddle horse—left the old timers, saying I would be back in a week's time, and hoping that by then they would have found the mine.

Arriving at Duran, my partner and I went into a huddle, and discussed the fors and againsts of the whole situation. The fors were the legend concerning the existence of a rich deposit of gold, known to most every Mexican family from the Pecos River to the Rio Grande. In repeating the story many of the little details did not agree, but the main fact that a deposit of gold had at one time been found was agreed to by all. Then the facts of the brother from Bernalillo, having worked two years in the Pedernals to discover the vein, was another in his favor, and Colonel Hopewell's, and my several assays, both running over a hundred dollars in gold, yes, decidedly we had the mine in sight. The weight of evidence was largely in favor of finding the mine; in fact no argument could we offer against success.

A week later I rode to the Pedernal camp, and not finding the boys there, and concluding they would be back presently, sat down to wait.

A little before sundown they arrived, each with a sack of samples on his back. "Well, boys! What luck!" I offered by way of greeting.

"Well," they replied, "we must have over five hundred pounds of samples, all tagged showing where they were found, but there will have to be a lot of assaying done to determine if they are of any value."

I loaded the sacks into the buck-board and left for the railroad, shipping them by express to El Paso, and calling for an immediate assay, for gold only.

Four days later I received the returns, not a trace of gold in any sample submitted, though many of them looked similar to the sample from which I got returns of over one hundred dollars to the ton.

Upon my return to the camp, I looked for the boys to be bitterly disappointed, but during my absence they had met the two Colorado prospectors, who told them that out of fifty samples they had sent to Mr. Hopewell, not a one showed a trace of gold. We worked a month longer, and quit. Somewhere in those Pedernal hills undoubtedly is a rich lead of gold. Who will be lucky enough to find it?

The Dead Burro Mine

by N. Howard Thorp

It is an old story, one of the many myths of New Mexico, the tale of the Dead Burro Mine.

Many years ago, I first heard the story told to me by an old Mexican, living at La Jolla on the Rio Grande.

I had had a hard day's ride, and my saddle horse and I noticed a little smoke arising from a clump of cottonwoods along the river at about the same time, and his ears working back and forth, soon indicated a small adobe cabin ahead.

Approaching the door, the usual greetings passed between Manuel Torrero the owner of the cabin and myself, and invited in I sat down in front of a cheerful fire in an adobe fireplace. It is unnecessary to enter into a description of the little adobe cabin, as it was a replica of thousands of others, whose owners are of the poorer class of New Mexico rancheros.

Riding up the Rio Grande Valley that afternoon, I had seen a dead burro, and in the course of conversation, after supper, remarked concerning it, and the unusual fact of a burro ever dying, except by accident, evoked the following story, rumors of which I had frequently heard, of the Dead Burro Mine.

Some sixty miles east of where we were then sitting, and after passing the Organ Mountains, the white plains, and salt flats, one comes to the southern end of the Guadalupe Mountains, from whose eastern side flow several streams, but whose western side approaches are devoid of water. This flat to the west of the mountains was where the famous salt war had its origin, and was the cause of so many being killed in the old town of Ysleta, Texas.

My host said that as a boy he knew an Indian named Juan Reyes who at that time was considered very rich, and had a gold mine which no one else could find.

At that time Manuel Torrero was a sheep-herder for a member of the Saiz family, who ran many flocks of sheep, and whose range was from the Rio Grande River east, and during the rainy season when the surface lakes were full of water, their flocks would drift east as far as the Guadalupe Mountains, then as the lakes dried up, drift back west to their home range along the Rio Grande River, where the flocks were snared and lambed, and where was situated the Hacienda of his patron Don Antonio Saiz.

Manuel told me that many people tried to follow Juan Reyes to find out where his gold mine was located, but 'twas of no avail. Many times, Manuel said, when Juan Reyes was watched so closely he could not leave, he would give a baile, and invite all the people.

When everyone was enjoying themselves, Reyes would slip away on his famous burro "Sal" and be gone for a month, but when again showing up in the settlement, he would always have plenty of money.

The rainy season having arrived, Manuel and his flock had been started grazing east, so as to get the benefit of the new grass and water. Camped one night on what is called La Seja de Alamo, he noticed the fresh tracks of a burro, evidently carrying a man or load, headed in the same direction his herd was going. It was not the tracks of a loose burro grazing along, but one—from the straight direction he held—controlled by a rider. Manuel followed the trail for over a mile, the tracks bearing directly for the point of the Guadalupe Mountains, some thirty miles to the east. The burro of Juan Reyes was as large as a Mexican Mule, in fact, weighed over six-hundred pounds, had speed, and lots of endurance, and although Reyes owned plenty of saddle horses, and other saddle stock, when he disappeared from home, Sal always was his mount. Juan Reyes, like all Vaqueros, rode a rather gaudy saddle, and on the back of the cantle could plainly be read a large J. R. outlined by silver tacks driven into the tree of the saddle. The probable reason for the use of a burro on his expeditions was the fact that his mine lay in a dry portion of the country, where a horse would suffer from a lack of water, but a burro could maintain life and thrive on the various succulent desert forms of cactus.

Then again a burro was not so easily tracked, and his tough

hooves could stand rocks, where an unshod horse would suffer and soon go lame.

Years passed, until early in the eighties the Apache Indians' raids got so bad, so many settlements being attacked and herders killed, and their flocks stolen, that the different patrons kept their herds close to their ranches in the Rio Grande Valley.

During this Indian excitement, Juan Reyes seemed to bear a charmed life, going and coming as he pleased. No attention much was paid to him until one day someone asked someone else if they had seen Reyes. No one could remember having seen him under two months. In fact, he never returned.

A few years after Reyes' disappearance, Manuel had accumulated some cows, and had quit herding sheep for his old patron.

Late one fall, Manuel and his boy Giorge started for the Guadalupe Mountains to hunt as was their yearly custom. As in those days the Guadalupe Mountains were alive with elk, Manuel and his son each riding their horses drove ahead of them a bunch of burros, some loaded with water kegs, bedding and camp outfit, and another dozen or so to carry back the dried elk meat after they had jerked it for their winter's supply.

After several days' journey they made camp at the point of the mountain, and at the foot of the big bluff, with its natural cross, and piled the outfit into a cave.

This cave is known as Cueva de Oso, or Bear Cave, on account of its use by bears, and which long afterwards was the hide-out of the famous outlaw Singleton. Some distance north and west of the point of the mountain, a portion of it had split off, leaving a narrow chasm or canyon, running north and south, a hundred feet or so in width, and a half mile or so in length. At the head of this canyon, there is a deep pool of water, caused by whatever rain there is running down the sides of the adjoining cliffs. As it is accessible only to mountain goats, eagles, and a man with a rope and bucket, this pool—though in times of drought becoming slightly stagnant—always contains water, and Manuel and his son were two of the few men in New Mexico who knew of its existence.

In a small canyon adjoining the main one, and which though

quite long was very narrow, Manuel came across the dried remains of a man and burro lying almost side by side. The burro was identified by the saddle as belonging to Juan Reyes' "Old Sal" who had been out with him on so many gold hunting expeditions. The body of the man was Juan Reyes, who had been scalped, and both upon close examination proved to have been shot through the head.

The small maletas on the burro's saddle contained a sack of very coarse gold dust, mostly nuggets, which later proved to have a value of about twelve hundred dollars.

It was evident that Juan Reyes had completed his gold gathering and was starting home when surprised and killed by the Indians. Manuel and his son buried Juan Reyes as best they could, carrying the burro's saddle and gold back to their camp. As the narrow canyon was almost always in the shade, the saddle was still in fair condition. Manuel, although now a man of middle age, with his son Giorge made yearly trips in search of the gold. Although he claims never to have found it, the source from where the gold came, such persistence should not go unrewarded, and it is not improbable that should Manuel not live long enough to attain this end, one of his grandsons may someday run across the Dead Burro Mine.

The End

Mrs. Frances Totty
Box 677
Silver City, N. M.

Date: Nove. 6, 1937
Words: 315
Subject: A discovery of a Cave

A Discovery of a Cave

Roscoe Rodgers lived on a ranch at Lone Mountain with his father Clark Rodgers.

One day while hunting he noticed a small opening among some rocks After pulling away some shubbery and stones a good sized opening was revealed which when he went into the cave he discovered relics of days gone by.

The floor was covered with a varied assortment of Spanish and Indian relics, among the relics were several old Spanish saddles, bridles, bits and augers upon which were engraved the names of their Spanish makers; a number of bows and arrows, the arrows still retained their steel points which were covered with a red rust which showed that they had served their owner only too well.

A powder horn was also among the relics, being in a good state of preservation. The stirrups of the saddles were of the finest carved wood, and well preserved while the leather was badly decayed, but beautifully carved.

The entire riding outfit probably belonged to some Spanish officer.

Laying along-side of the relics and stretched at full length on the floor was a human skelton, which was judged from the high cheek bones, was undoubtly an Indian, supposed to be a peon of the Spanish in the Indian Uprising in 1860, when they attacked the Spanish mines at Santa Rita, and caused them to erect the rude adobe fort, portions of which can be still be seen at that place.

Buried Money on the Mimbres

by Frances E. Totty

Pedro Raesequeon came to Grant County in 1880 from Mexico and settled on the Mimbres River near the present site of the Mimbres post office. Mr. Raesequeon was a freighter by trade and always freighted into Grant County to the various points of importance from the lower Rio Grande valley and Mexico. Mr. Raesequeon when he came to the Mimbres was about thirty years old. He and his wife started a small ranch which she took care of while he was away from home freighting. The family lived at home entirely, and when it became necessary to buy any groceries at the store they never bought them, but traded eggs, butter or other products for their meager supplies. If the family ever spent any money no one ever knew about it, unless Mr. Raesequeon spent some money on his trips to Mexico and to the lower Rio Grande. They had the barest necessities of life, but it was a known fact that they were making money from the freighter line as well as on the small ranch they owned for they were a group of thrifty people.

Mr. Raesequeon in 1918 sold a part of his ranch for $17,000 cash, but was not known to place the money in the bank, and his son, Manuel Raesequeon, Mimbres, New Mexico, says that he is positive that his father buried the money along with all that he had acquired during his long period of freighting to Grant County's various small towns of the early days. A small part of the money was found at one time in 1935 it is believed, but it might have been some that was buried by someone else. This money was found by Manuel.

The family still has ranch holdings in Mexico. Pedro Raesequeon died in 1935 at the age of 85. Some of his buried money may someday be recovered, but it is doubtful.

Just across the road from the Raesequeon place is the old Ancheta place where Louis Ancheta moved to when he left Pinos Altos to settle on the Mimbres. Louis Ancheta came to Pinos Altos around 1860, after having escaped from Old Mexico after having taken part in a revolution. During the revolution Louis fatally wounded his father, who before his death placed a curse on his son, Louis Ancheta, the curse being that Louis would marry and have a large family and that every child would disgrace the Ancheta name, and that Louis would acquire a large fortune and would not ever enjoy the pleasures that money could give.

When Louis Ancheta settled in Pinos Altos there were only a few settlers there at the time—in a short time the place was flooded with miners and prospectors all in search of gold. Louis Ancheta soon found that he could acquire a fortune by buying the virgin gold. Louis Ancheta married the daughter of a rancher near Pinos Altos, and a short time after their marriage they decided to move to the Mimbres River near the present site of the Mimbres post office.

The gold was moved from the Ancheta home in Pinos Altos to the Mimbres in large champagne boxes that were held together with screws. Robert Stevens, a youngster at the time living in Pinos Altos, said that all of the boxes were sealed but one, and that it was only partially filled. After he was grown and knew the value of gold and weights he estimated that there was probably $150,000 worth of bullion stored in the boxes.

The Ancheta family all seemed to be under the curse of their grandfather and some came to meet a tragic end. Juan Ancheta, San Lorenza, New Mexico, grandson of Louis Ancheta, will allow anyone to dig for the bullion which his grandfather buried, but he isn't interested in searching for the bullion himself.

Source of Information: Primie Isleta was born and raised in Santa Rita—born 1888.

Hidden Treasure

by Mrs. Frances Totty

Colonel Van Potten, register of the Las Cruces Land Office of many years ago, one of the pioneers of this country and very influential among the Indian and Mexican residents of the Mesilla Valley, heard of an aged Indian woman living in the Mesilla Valley who was telling a strange tale of a treasure trove hidden in the mountains near Santa Rita. Mr. Van Potten went to the woman to verify the tale.

The old squaw told him a harrowing tale of Jhu, a chief of the Apaches in the early days and his braves holding up an ore train in northern Mexico, bound from some of the rich mines of the Yaqui district to Chihuahua, the capitol of the State.

Bullion to the extent of the $40,000 was obtained and Jhu, after satisfying the braves with a share of the plunder, took the $40,000 in bullion and forbidding any of his men to follow him, took five squaws with him and rode north as fast as he could.

After crossing the Gila River, he rode up a broad canyon which at the time was a favorite rendezvous of Mangas Colorado and other chiefs, and is easily recognized today as Greenwood Canyon, thirty miles northwest of Silver City.

Here the squaw said he deposited the billion beneath one of the overhanging cliffs with which the canyon is bordered and then he had the squaws seal the cliff. This done, Jhu took his war club and beat the five squaws in the head.

The old woman was left for dead with her four companions, but after laying unconscious for several hours, she came to, and found aid among the friendly Apaches that infested this section of the country at that time.

After Geronimo's capture she settled in Santa Rita and from there drifted to the Mesilla Valley where some of the descendants of her tribe lived. The old woman could never be induced to go back to the scene of the hidden treasure for fear she would be captured by the enraged Jhu, or some of his descendants, for revealing the tale.

There are many caves in Greenwood Canyon, but it's known there are caves where Indians dwelled and several sealed caves have been opened; but the cave with the $40,000 of bullion has never been opened as yet.

Source of Information: Gene Heron.

F. Totty   The Story

SEP 22 1937

"I befriended a Spanish fellow who told me to go 32 miles North of Silver City, and gave locations. There he said I would find hidden by nature in a cave a bell bearing the Spanish Coate of Arms, Aramors, Skeltons, and other relics either put there by Indians or the Spanish became lost and remained to die in this cave.

"This bell is very large, and is probably worth money"I was told.

"Whose land is this cave on"? I asked.

"Government ", was the reply.

"Why don't you turn this information over to someone to investigate, who is an authority,".

"Why should I"?

"Because it is a thing of intrest to the people, the government, and if authentic would be of a historical value to our government both federal, and state."

"Would I get any thing if I turned over my information?"

"You would probably if it was of any value. I have always found our Government fair to those that have anything of value to the country."

This party stated there was a map of the Southwest Carved in the rock at this cave with all rivers, creeks and mountains indicated.

He claims he has pictures of the place also a small goddess made of silver that looks Egyptian that he took from the cave.

Louis Ancheta

by Frances E. Totty

Louis Ancheta came to Grant county in 1860 after having taken part in a revolution in Old Mexico. Louis was on the side opposite to his father, and during a conflict with the other faction Louis shot and fatally wounded his father. Before his father died he placed a curse upon his son that had mortally wounded him. The curse being that Louis would marry and have a large family, and acquire wealth, but for some reason this wealth would do no one any good.

Louis came to Pinos Altos a short time after Snively, Birch, and Hicks discovered gold at Pinos Altos in 1860. The gold was first discovered by Birch as he was taking a drink out of Bear Creek just above the junction with Little Cherry. Louis began to acquire all the gold in the nearby district. He bought the bullion for some time, and asked why he didn't place the money in the bank he remarked that he had more gold than any bank in this district and why should he place it in the banks.

In the early '80's Mr. Ancheta decided to move over on the Mimbres at the present site of Mimbres Post Office. The bullion was placed in the champagne boxes. The boxes of the early days were made very strong as everything was shipped by ox team from Las Vegas. The corners were screwed together instead of being nailed.

Robert Stevens and a number of children were playing around the wagon when they were loading the bullion to move it over on the river. There was one box that was only partly filled with the bullion, but the rest were filled and closed. It has been estimated that there was probably $150,000.00 at the very least. Mr. Ancheta moved this bullion to Mimbres, some 40 miles northeast of Silver City. Robert Stevens said that he had often wondered how pure the bullion was; he looked into the box that was partly filled and has often remarked he never did see any better specimen of ore melted into bullion.

After the Ancheta family had been in Mimbres for some time they decided to move the ore from its hiding place behind the piano in the front room. Mr. Ancheta had a man come and dig a pit in the back yard and carry the bullion out in a wheel barrow and place it in the pit. For some reason Mr. Ancheta seemed to be afraid that the bullion was going to be stolen, and decided to move it again. Where the bullion was moved to by Mr. Ancheta is still a mystery. There had been in recent years searches made for the gold. The curse placed on Louis Ancheta was felt to the fullest extent. Mr. Ancheta buried the gold bullion, and a short time later was killed. His sons all came to a tragic death. One committed suicide in Silver City, another was killed over on the Mimbres river in a drunken brawl. The queer thing is that the children did disgrace the Ancheta name, but the grandchildren are of the very best of American Citizens and are a highly thought of family of people. It has been suggested from time to time that Louis Ancheta might have sent the gold bullion back to Old Mexico to help out in revolutions, but if he did there isn't any way to find out as he was very secretive about his affairs, but if he did he surely sent it out secretly, because there wasn't any of the children that knew where it was hidden the last time, and spent many fruitless hours trying to find the buried money.

Today the grandson of Louis Ancheta allows people to dig for the bullion, and the Spanish people of this district still think someday they will find the bullion, but the curse that rests on the family has been so keenly felt that the Ancheta family doesn't care too much for the money. Every child of the Louis Ancheta family came to a tragic end, even to the girls.

Source of Information: Mike Houghes settled in Grant Co. in 1865. Mike Houghes' family came to Grant County in 1865 from Ireland; the family had been one of the largest ranch holders in the county at one time. Mike Houghes is the last surviving member of the family, and still lives out on the original ranch of the family on the Mimbres river. Nearly all of the ranch holdings of the family have gone, but in this old home one finds the true western spirit. Mike Houghes' father was one of the first commissioners of Grant County, and the family has always taken an interest in the development of the county.

The Ghost of Georgetown

by Mrs. W. Totty

John Barachman was the first settler of Georgetown, which was settled in ----, which grew to be quite a city.

Georgetown at one time had approximately twelve thousand people, two banks, eighteen saloons, and one newspaper. For many years it was thought Georgetown would be selected as the county seat of Grant County, but as time went on and the mines closed down, Georgetown's population began to decrease. This one-time thriving little city is now a place of a dozen families.

There is a small Catholic Church still standing in this former Mecca that was dedicated as the Sacred Hearts Church with its first Sexton being Trinidad Barela.

Trinidad Barela and Governor Larzola were protégés of Archbishop Borgada.

Trinidad Barela was the first person known to tell the story of Onecimo Archuleta to the natives of this district, that have handed down the story by mouth.

Onecimo Archuleta was a very wealthy man sent by the Mexican Government into this country many years before the Gadsden Purchase.

Onecimo Archuleta told his followers that when he died he wished to be buried in Georgetown, which would be a very rich gold country. Georgetown at this time had never been founded.

Onecimo Archuleta settled in what is now Santa Rita, thinking he had reached Georgetown. He built a wall, which part of still stands behind the old Santa Rita Warehouse. Archuleta Wall was to be around his home to guard his gold from the Indians, but he discovered too late that he was some few miles south of his destination.

Onecimo Archuleta was burned to death at Santa Rita by his

followers, who were hoping to get his wealth, but they were never able to find Georgetown or Archuleta's mines after his death.

Archuleta's ghost comes to Georgetown and walks in the Sacred Heart Church. The Spanish people say he is walking in the Church because he was not buried at Georgetown as he wished. So his spirit cannot rest until his wish is carried out.

Source of Information: Miss Trall.

Mrs. W.C. Totty
Box 677
Silver City, N.M.

Words: 308
Date: Nov. 20, 193'

Lost Treausres

There is an old tradition in this section of the country that many years ago when Ft. Bayard was in its days of youth, and perhaps before the founding of San Vicenti, now Silver City, a military pay-wagon was held up by masked bandits a few miles east of the present site of Silver City on the old Santa Fe trail that led from Santa Fe to Tucson and Fort Huachaca in Arizona.

Many thousands of dollars of the government money intended to pay off the soldiers at the Arizona post were taken, and to avoid detection, was cached beside the road.

An old Mexican woman who lived in Santa Rita claimed to have a record of the hiding place of the hidden treasure. This record being left to her by descendants of one of the bandits.

This woman gave this rocord to two young men Ybarra and Navarez, by name, who hunted for the treasure. They were instructed to follow the old Military road east from Silver City for a couple of miles until they came to a cross made of stones beside the road.

This cross supposedly marked the place the money was cached. The men told that they easily located the cross and proceeded to dig for the treasure.

The men found a number of ancient Spanish coins of small demomination, but that was all. The great treasure has never been found yet as far as is known.

Informant: Mrs. Gene Fleming.

The Ghost of Priors Canyon

By Frances E. Totty

Many years ago, sometime in the early sixties, a prospector settled in the northwestern part of Grant County, some fifty miles from Silver City, in what is now known as Priors Canyon. As to what the prospector's name was no one ever knew, and no one ever questioned him in any way for in those days everyone's business was his own, and anyone that asked too many questions was likely to be buried with their boots on.

    This prospector built a log cabin and settled in the canyon going around in the Black Range and the Mogollon Mountains prospecting. He always had gold and was said to have staked out a good claim someplace in the mountains. It was thought that he was burying his gold in the cabin or around it. One day, a traveler went by the cabin to spend the night, as in those days houses were far apart, and found the old Prospector dead and the cabin raided, but not by Indians, and it has always been thought that he was killed for the gold that he had in the cabin.

    A short time after this another traveler was passing by the house and decided to stay for the night and after looking around he saw that there wasn't anyone around and started to fix his evening meal when a man appeared around the corner of the house, dressed in a frock tail coat, and a smoke stack hat; he had a mustache, and was cleanly shaved. The traveler spoke and asked him to stay for the evening meal; the man turned and walked back around the house, and when the evening meal was ready to eat the traveler went around the house to call the stranger to eat, and could not find him; he called and called and didn't get an answer; after some time he started hunting tracks of the man and discovered that there wasn't any. The traveler then went into the cabin to sleep for the night, and some time after he had gone to sleep the cover was pulled off of him as if a large hand had reached down and lifted it off of his body; he reached up and got the cover and wrapped it around his body and covering his head the cover was again lifted from his body. After the second time his cover was raised up he decided that the house wasn't any place for him. He got up, took his bedding and slept out near a tree and didn't have any more trouble. The traveler told his story after he left the canyon, and everyone laughed at him.

Soon after this every traveler that went to the cabin had the same experience, and it was the talk of the county that the cabin of Priors Canyon was haunted. It has been said that the man doesn't appear anymore, but no one has ever been able to sleep in the cabin, for all of the cover is always pulled off. Among those that have been to the cabin are Ed Ward, Silver City, New Mexico and Bob Stevens, a trapper.

They say that you can go to bed and you will naturally be expecting something to happen and will not go to sleep very easily, but when you do you will be awakened by the covers being lifted off, and when you get good and awake the covers will be off and lying on the floor at the foot of the bed and you can get up and wrap yourself up in them and it will seem that your skin will be pulled off in the effort of the covers being pulled from around your body.

For many years there has been a light appearing on the ridge between here, Mimbres, and Santa Rita about midway. This light does not seem to burn out but goes out and then comes back on. I have seen the light and after seeing it have gone up on the ridge and searched for its source, but have never found any. Many people have seen the light. The last person to see it was Henry Aklin at Aklin's store, some month or so ago. I know that the light has been seen for the last fifty years, and have heard my father say it was there before then.

Mike Houghes was born on the Mimbres River at the present home site. His parents came here direct from Ireland. His father, Nick Houghes, was commissioner of Grant County in the early days and his stepfather, Alec McGregor, was an important figure in the government of Grant County.

Editor's Notes: The prospector's cabin still stands. Store located at foot of hill 12 miles northeast of Santa Rita and 30 miles from Silver City.

The hill is called the Silver Lodge by old timers. Alec McGregor was first commissioner of Grant County after Silver City was established in 1873.

Source of Information: Mike Houghes, common story on Mimbres, born 1887 on Mimbres.

Wilcox Mining Claim

by Frances E. Totty

The day that the Apaches attacked the Roberts' cabin, Mr. Wilcox, a miner, came to the house hunting his partner. When the news came that the Indians were out, the man remained at the cabin. He showed the men a sample of ore he had brought back from the mountains. The nuggets were pure gold and unusually large. Mr. Wilcox had gone into the mountains alone. When he discovered the ore he came back to the mining camp after his partner. Mr. Wilcox told the men, "There is enough gold where this came from for my partner, myself and the entire settlement. When we go back the men can go file on claims, as it isn't far."

Later in the day Mr. Wilcox saw his partner trying to come to the cabin; he knew that the man was in a dangerous place. Mr. Wilcox made an effort to save the man's life thereby losing his own. Mr. Wilcox stepped to the door to tell the man the best way to get to the house. When Mr. Wilcox showed himself at the door, Mr. Indian saw a chance to add another white man to his list of murders of the frontiersman. With the words, "My God, boys, I'm shot," Mr. Wilcox, who stood by the door with his gun, walked over to the fireplace and laid down on the hearth. When they got to him, he was dead.

With the killing of Mr. Wilcox went the secret of where the ore was to be found. There wasn't anyone that knew the direction the man had come from when he arrived at the cabin. It was thought that the claim wasn't far, as the man had told them that it wasn't, but the claim was never discovered.

Many trips were made hunting the claim, but all were in vain as not a tool of any description was ever found. Mr. Wilcox must have buried his tools and covered up all signs of his prospecting for if the place was ever found it was not known.

Source of Information: Agnes Meader Snyder.

According to legend, Indian slave labor was used in the old Spanish mines of New Mexico. Actually, very little mining was done in those days. Simple ladders made of notched logs were used in Indian pueblos and, later, in pit mines. Ritch, 1885, *New Mexico in the 19th Century: A Pictorial History,* Andrew K. Gregg, p. 171.

Mexican Arrastra. *Illustrated New Mexico 1885*, W. G. Ritch, Fifth Edition.

The Placer Mines—working the Rocker, New Mexico (?), Charles F. Lummis, NMHM/DCA #101882.

Stoping a mine. This method is not used in a strip mine such as Santa Rita. Thayer, 1888, *New Mexico in the 19th Century: A Pictorial History,* Andrew K. Gregg, p. 172.

Lt. Emory's detachment, on reconnoissance, were among the Anglos to pay an official visit to the Santa Rita mines. Emory, 1848, *New Mexico in the 19th Century: A Pictorial History,* Andrew K. Gregg, p. 173.

Unidentified miners near Silver City, New Mexico, ca 1889-92. Rev. Ruben Edward Pierce, NMHM/DCA #093812.

Central Mine from the W near Silver City, New Mexico. J. R. Riddle, NMHM/DCA #076118.

"U.S. Treasure" mine and mill, Chloride, New Mexico, ca. 1890. Henry A. Schmidt, NMHM/DCA #012633.

Dr. Haskell's cabinet of mineral specimens, Chloride, New Mexico, February 17, 1883. Henry A. Schmidt, NMHM/DCA #065570.

The town of Lake Valley, west of Hatch, is one of several old mining camps in that area that are now ghost towns. *New Mexico in the 19th Century: A Pictorial History*, Andrew K. Gregg, 133.

"Bridal Chamber Mine," Lake Valley, New Mexico, ca. 1890. Henry A. Schmidt, NMHM/DCA #012655.

Group of Miners, "Bridal Chamber Mine," Lake Valley, New Mexico, ca 1890-1895, NMHM/DCA #056218.

Lode mining works veins of ore to be processed above ground. Thayer, 1888, *New Mexico in the 19th Century: A Pictorial History,* Andrew K. Gregg, p. 173.

Smelting Works at Lake Valley. The Bridal Chamber mine there once produced $3 million in horn silver ore in six months. Thayer, 1888, *New Mexico in the 19th Century: A Pictorial History,* Andrew K. Gregg, p. 133.

Stamp Mill, White Oaks or Fort Stanton 1893. Left to Right: 1. Will Lane, 2. I. N. Bailey, 3. Dave Girdwood, 4. Will Real (Lincoln Co.). NMHM/DCA #089680.

Miners in Lincoln Country, New Mexico, 1904. NMHM/DCA #005243.

"Struck It Big," North Home Stake Mine, White Oaks, New Mexico. NMHM/DCA #163580.

Three unidentified miners, Raton/Yankee, New Mexico area. NMHM/DCA #091121.

Entrance to coal mine, Raton-Yankee area, New Mexico, 1900? NMHM/DCA #091138.

Raton Coal and Coke Company Coal Mines near Blossburg, NM, ca 1893. NMHM/DCA #014257.

Unidentified Group of Miners, New Mexico 1890? NMHM/DCA #112985.

Pacific Stamp Mill, Pinos Altos, New Mexico, 1890? L. A. Skelly, NMHM/DCA #148540.

Ore was crushed in stamp mills such as this. A drive wheel lifted the vertical hammer rods, letting them fall to pulverize the rock. The thunder of these old mills rumbled across the valleys, and the foundations of these mills still can be seen in abandoned workings. Bell, 1870, *New Mexico in the 19th Century: A Pictorial History,* Andrew K. Gregg, p. 171.

Panning gold in the mountains near Cerrillos. Gregg, 1849, *New Mexico in the 19th Century: A Pictorial History,* Andrew K. Gregg, p. 173.

Mrs. Captain Jack, Mining Queen of the Rocky Mountains, 1910? NMHM/DCA #156599.

Ruelina Camp and Shaft, Cerrillos, New Mexico, ca 1881. George C. Bennett, NMHM/DCA #014840.

Homestake Shaft, Cerrillos, NM, ca 1881. George C. Bennett, NMHM/DCA #014843.

Entrance to the Grant Tunnel, Chalchihuitl Mine, Cerrilllos, NM, ca 1881, George C. Bennett, NMHM/DCA #014826.

Blowing with Washer, Golden, New Mexico. NMHM/DCA #154787.

21 oz. 5 p.n placer Gold. Result of the operations of The Santa Fe Dredging Co., Golden, New Mexico Parkhurst, NMHM/DCA #012647.

Prospector and burros in front of photograph gallery, Hillsboro, Kingston, New Mexico, ca 1892-1900, George T. Miller, NMHM/DCA #076511.

Panning gold near Hillsboro. (?) Davis, Henry A. Schmidt, NMHM/DCA #148168.

Man with pickaxe, mining area of Hillsboro and Kingston, New Mexico, 1885-1892?
Henry A. Schmidt, NMHM/DCA #076432.

No. 1 Tunnel of Good Hope Mine, Hillsboro, NM, ca. 1890. Henry A. Schmidt, NMHM/DCA #065124.

Lady Franklin Mine, Bullion Hill, Kingston, New Mexico, NMHM/DCA #160404.

Mrs. Sadie Orchard on right, in front of Ocean Grove Hotel, Hillsboro, New Mexico 1895-1902. George T. Miller, NMHM/DCA #076560.

Chinese laborer, mining area of Hillsboro and Kingston, New Mexico, 1885-1886? NMHM/DCA #076201.

"Elegantly dressed negro woman (possibly one of Sadie Orchard's girls), mining area of Hillsboro and Kingston, New Mexico, 1885-1892? J. C. Burge, NMHM/DCA #076431.

Malachite Bill, miner Albright Parlors, Albuquerque, New Mexico, 1890, NMHM/DCA #090392.

School of Mines, Socorro, *New Mexico*, Max Frost, 1894.

PART TWO: OLD MINES

"'If you kill me,' I said, 'my partner will be back and see that you hang for it.' 'I'll fix your partner the same way, you claim-jumping cur.'"

—From "A Prospector's Experience"
by Mrs. W. C. Totty

Humorous Incidents of Early Mining Days: A False Alarm

by James A. Burns

The construction industry generally and placer mining particularly, have in the course of time developed a code of steam whistle signals, which have become general and universally recognized in the industry. One long, one short blast—start work; two long, one short—shut down; three long—water wanted, start the pumps; four long—Boss Call, break down or accident; five long—Fire! Everyone turn out to fight fire!

Where no steam plant is in operation, the alarm is given by fire rifle or revolver shots in rapid succession. The giving of a false alarm, especially for fire, is punishable by immediate discharge from the employ of the company.

Such an incident, where just punishment was averted from the guilty parties, happened at the workings of the Moreno Gold Gravel Company at their camp on Six Mile Creek, a mile or so northwest of E-town in western Colfax County, in the Fall of 1890. E-town had passed the boom days of the 70's but there was still considerable placer mining on the creeks in the Moreno Valley and the miners spent most of their idle time and money in the saloons and other resorts of the village.

Late in the fall of 1890 it began to get pretty cold nights at the camp of the Moreno Gold and Copper Company, so the superintendent E. L. Hall put on a night foreman George Kelly to keep up steam and keep the pumps from freezing.

One cold night Kelly and Johnson the night watchmen at the sluice got tired sitting around the fire room and crawled up on top of the boilers and went to sleep.

Along toward morning as the fire burned down and the fire room cooled down to about the outside temperature they both woke up.

Said George, "Say Ed, is the fire all out? Have we any steam left?"

"How the hell do I know?"

"Well, try the whistle and find out."

Johnson jumped to the floor, grabbed the whistle cord, and pulled down for one long blast and then, excited by his success, blew four more long ones, making five long ones, the universal fire alarm.

They could see a dozen lights come to life in the camp and several in the neighboring village of E-town. Realizing his mistake Johnson said:

"Now we have done it! What the hell will we do?"

Said Kelly, "Begorra, I know, we'll both go out looking for the fire! You go up the creek and I'll go down toward camp; if you meet anybody ask them where the fire is."

So they separated and left before the rush of rudely awakened miners.

They kept off the trails, circled around and returned to the pump house. They found the superintendent and a bunch of miners, mostly half dressed and packing guns of every description.

They began asking, "Where is the fire?" "Who blew the alarm?" etc.

The miners began asking one another, "Did you hear the alarm, Mr. Hall?" "No," said the superintendent "but Mr. Carrington (the bookkeeper) told me he heard it." And so on, everyone including the two guilty birds swearing that they had not heard the alarm. Finally they found a Spanish American boy, Pedro Archuleta, who admitted he had heard the alarm.

"You are the only one that heard it! You must be the *hombre* who blew the whistle, you deserve a good swift kick." Which they proceeded to give him, and several of them, and then they started back to camp, grumbling their disgust over their broken rest.

The two guilty rascals squared their consciences by buying the poor Spanish boy all the whiskey he could drink the next time they met him in E-town, and so the incident passed into history.

Source of Information: Personal experiences of H. E. Anderson, Raton, as told to the writer. Names fictitious.

Humorous Incidents of Early Mining Days: A Self Made Reputation

by James A. Burns

In all the countries and especially in frontier districts, the seriousness of crime, and the punishment due for law violation, is measured by ways of living and working of the people in the community generally.

In the East, the stealing of a farmer's horse was only a little above petty larceny, just a little more serious than chicken stealing. In the old West, among the cattlemen, it became the one unpardonable sin, punishable by death by hanging if caught. Not so much for the value of the horse, or even the forty dollar saddle on the ten dollar horse, but because the loss of a horse, forty miles from nowhere, put the cowboy in danger of death from starvation and more particularly thirst before reaching a habitation.

Among the gold miners, especially the placer miners, the most detested crime, and warranting instant death if caught in the act, was the stealing of Amalgam (quick silver and gold) from the riffles in the sluice boxes. These were of necessity left exposed in the open until the regular weekly or monthly clean-up, when the Amalgam was gathered from the riffles, the quick returned to the riffles, and the gold, the miners' harvest, shipped to the mint.

Where the placer workings were located near small mining towns, far from law and order, it became necessary to employ watchmen, usually trusted employees deputized by the sheriff, to patrol the line of sluice boxes from dark until daylight. With some of those quick-triggered watchmen, it was dangerous for anyone, even Superintendents or owners, to approach the line of sluice boxes during the night. Many of those deputies worked on the theory "shoot first and investigate afterwards."

How a young miner took advantage of this situation, by act of canniness yet apparent dumbness—was related to the writer as follows:

In 1889 E. H. Johnson, then a young man, was night watchman for the Moreno Valley Gold and Gravel Company and their placer workings on Six Mile Creek north of E-town. The President of the company, Mr. Bloomer, had come up to watch the cleanup but they had not finished by quitting time, leaving about $4000 worth of Amalgam on the riffles with about 4 inches of water running through the sluice boxes.

The Superintendent, E. L. Hall, had given Johnson orders to arrest anyone coming near the sluice boxes; so buckling a six gun on his hip and calling his dog, who was a better watchman than he himself, he started making his rounds up and down the line. About 11 o'clock, as a late moon rose, his dog gave a growl and looking down the creek he saw Mr. Bloomer, the President, who had a habit of taking a stroll before bedtime, slowly making his way along the line of sluices.

So he let Mr. Bloomer approach a little closer. Then "Grab for the stars! Pronto!" Mr. Bloomer, startled, came to a sudden stop and saw himself covered by an old style Colt .45. He exclaimed, "Don't shoot, Johnson, it is me, Mr. Bloomer, don't you know me?" "I don't know anybody this time of night. You are under arrest, march." "My God, man, what are you going to do?" "I am going to put you where my dog won't bite you." So Johnson marched the president ahead of him to the camp office, a sturdily built log building, opened the door, shoved Mr. Bloomer in, and locked the door with the key and also the padlock used whenever the camp was temporarily abandoned.

Johnson whistled to the dog, "Come on, Perro, let's get back to work," and started back up the gulch until out of earshot of curses from the office.

He and the faithful dog resumed their lonely patrol of the sluice boxes without further incident, until daylight released them from their watch.

Coming to the boarding house for breakfast he met the superintendent, E. L. Hall.

"Good morning, Ed, did anything happen during the night?"

"Nothing much, I arrested one man about 11 o'clock for prowling around the sluice boxes." "The H--- you did, did you know him?" "O yeah, knew him as well as I know you, it was Mr. Bloomer." "My God! Didn't you know any better than that?" "You said to arrest anyone coming near the sluice boxes."

"Well, we'll see what we can do about it. Where did you put him, in jail?" "No, I didn't want to walk that far, and had to get back to work, so I locked him in the camp office."

So the Superintendent and Johnson walked back to the office and Hall unlocked the door, then noticed the padlock was still on.

"What were you afraid of, that someone would break in? Give me the other key."

They unlocked and finally opened the door and found Mr. Bloomer, with his clothes somewhat mussed up and dirty, but otherwise no worse for wear.

Said Hall, "How do you feel, Mr. Bloomer?" "O, not so worse, those planks in that empty bunk are not so dammed soft." Then catching sight of Johnson behind Hall, "Say, that's a dammed faithful watchman you have got but he takes his orders too dammed literally. Say, has the cook anything left for breakfast?" "If he has not, we'll have Ed here cook you up something." "All right" (dubiously), "but be sure and tell him not to put in any rat poison or something."

That was the last Johnson ever heard of the incident, except for good natured kidding by Hall and his fellow miners. But his job was good as long as he wanted to stay. He had "made himself a reputation" with the company.

Source of Information: Personal experiences of Harry E. Anderson, Raton, as told to the writer.

SEP 29 1931

FIRST MINE REGISTRATION IN NEW MEXICO
1685

First known registration of a mine in New Mexico - 1685.

At the pueblo of Nuestra Senora de Guadalupe de El Paso on the 26th day of March, 1685 - before Don Domingues Jironza Petris de Cruzate, Governor and Captain General, Pedro de Avalos, a soldier of the garrison of San Jose - Rio del Norte - claims a mine which is about forty-five leagues from the town and is located in the ridge of mountains and craggy rocks, which is called "Xotoreal" tower, (Fray Cristobal), that part of Rio Avajo del Norte to which mine I gave the name of Nra. Senora del Pilar de Zaragosa. He gave half of the mine to Captain Don Alonza Mael de Aguilar, and a part to his brother, Antonio de Abalos.

The property was discovered while on the campaign north with Cruzate for the recovery of the Province.

Justice Garcia de Marieza was called to verify the eastern part.

        Signed Pedro de Abalos
              (Avalos)

Approved by:
Signed Domingo Jironca
Also by Alplonizo del Ao.

From:
  Spanish Archive #1 - March 26, 1685.
  Spanish Archives, Vol. 1, page 1. Ralph Emerson Twitchell
  The Torch Press, Cedar Rapids, Iowa. 1914.

September 29, 193-, NMFWP, WPA #129, NMSRCA

Mines of Northern New Mexico
(Historical)

by James Burns

With the possible exception of some small scale placer mining by the Spanish American natives on what is known as the San Antone or Community grant near Valdez on the Rio Hondo in Taos County, the history of mining on a commercial scale in this district begins with the discovery of gold in the Moreno Valley in western Colfax County. The discovery of gold in this district is due to the discovery of copper by a Ute Indian on or near what is now known as Ute Creek on the east slope of Baldy Mountain. He exhibited samples of his findings to officers and others at old Fort Union. In the early fall of 1886, William Kreonig, W. H. Moore, and others sent three miners, Larry Bronson, --------- Kelly, and Peter Kinsinger, to do assessment work on a copper prospect on Baldy Mountain. These men arrived at Willow Creek and made camp. While his two partners were preparing supper, to kill time Kelly took a gold pan from their pack and began to pan the gravel on the edge of the creek; the first pan of gravel showed several colors of gold and when his companions heard his shout, they dropped their pots and skillets and all of them began to pan with whatever utensils were handy, obtaining gold from nearly every pan of gravel. Forgotten were their copper prospect and the assessment work they were sent out to do. They spent the balance of the fall digging trenches, pits, etc., until the early winter drove them out. They swore each other to secrecy, intending to return to Baldy Mountain in the following spring. But the temptation was irresistible; they showed some of the coarse gold they had brought back to Fort Union to some of their cronies at the Fort and from there the news spread to Las Vegas, Santa Fe, Trinidad, and other parts of New Mexico and Colorado. By the Spring of 1867 the rush had grown to a stampede and miners flashed in from all over the southwest and as far west as California. Bronson returned to Willow Creek with the earliest adventurers and staked out claims for himself and his partners. All claims were staked out up and down the creek from a big pine tree called "The Discovery Tree" which was still standing 40 years afterwards in the early years of this century and formed the basis for all later patent surveys in this district.

Among the prospectors who were attracted to the new field were several Irishmen and Cornish men who had gone round the Horn to California in 1850 or '51 and were familiar with the large scale methods of placer mining which had been successful in California. Among them were Matthew Lynch and Tim Foley. These men made locations for themselves on the south side of Willow Creek on the opposite side from the original discovery and then ranged further afield, prospecting around on the south and east sides of Baldy Mountain. Matt Lynch introduced hydraulic mining in New Mexico. He was successful with it, at times operating three or four machines each season and was one of the original discoverers of the Aztec Mine on the east slope of Baldy Mountain between Ute and Ponil Creeks.

Later in the season a later party from Fort Union consisting of Pat Lyons and associates arrived and, finding Willow Creek all taken up, moved north and west and discovered gold at Elizabethtown, more commonly known as E-town. As the Moreno Valley was more thoroughly prospected and more people flocked in, this settlement became the general headquarters for the district, and a town site was surveyed in the fall of 1879 and named Elizabethtown for Elizabeth Moore, daughter of John Moore, one of the prospectors. She afterwards married Joseph Lawrey, one of the prominent miners of the district and is said to be still living at E-town and owning mining property in the district.

Elizabethtown or E-town to use the common local name for it, which now has a population of eighty or a hundred and is a typical mining town of unpainted frame buildings—at one time had a population of eight thousand and was the first county seat of Colfax County and also the first town to be incorporated under the laws of the Territory of New Mexico.

Before this district was a year old, the experienced placer miners from California realized that they needed more water to work all the feasible projects. Someone, probably Matt Lynch, conceived the idea of bringing water from the upper branches of the Red River and delivering it at the upper end of the Moreno Valley workings. A fine idea; they had plenty of nerve and man power; all they lacked was engineering advice and money. For the engineering they went back to Fort Union and engaged the services of Captain N. S. Davis, an engineer officer of the United States Army. For the money they went to Lucien B. Maxwell, the patron of the Maxwell Land Grant on which the gold fields of the Moreno Valley are located. Maxwell was the big businessman of New Mexico, and seeing a chance for speculation, and possible profit from selling water, agreed to furnish the money. Captain Davis made the surveys in the fall of 1867 and construction work started in May, 1868. Employing about 400 men, the 48 miles of ditch, including side trill flumes, trestles etc., they took six months to build on an average 3-½ miles per month. Considering the roughness of the country and the distance from any base of supplies, this was a remarkable feat of engineering and executive ability. They practically built the entire ditch with few other tools than picks and shovels and double bitted axes. It is said to have cost a quarter million dollars including storage reservoirs on the West Fork of the Upper Red River. It is doubtful if the same ditch could be rebuilt for the same money today.

Place Names: Cities, Towns and Villages, Lincoln County

by Edith L. Crawford

White Oaks, Lincoln County, New Mexico T 6 S R 12 E. N. M. P. White Oaks, Lincoln County. Population 109. Altitude 6500 feet.

It was the first settlement on the west side of the mountains. Gold was discovered there in 1879, the "find" bringing many prospectors.

In 1880 the town was platted, lots sold, buildings erected and one of the most picturesque and substantial towns of the county was built. The town is in a valley with Baxter and Lone Mountains on the west and the Patos Mountains and Carrizo Peak on the southeast. It was named for the White Oaks Spring which is about two miles from White Oaks. The spring got its name from the white oak brush which grew around the spring. This spring was the only source of water that the early prospectors had and in the very early days the water was hauled in a water wagon. Later on it was run through pipes down to the town and even now you can trace this water line from the old rusted pieces of pipe where the line was. In 1880 a man by the name of Tom Williams discovered coal in the Patos Mountains about three miles from the town. Coal mines have been worked more or less extensively ever since.

The Lincoln County Light and Power Company owns a power plant in the mountains three miles from the town of White Oaks which furnishes power to White Oaks and Carrizozo, New Mexico. The population is about 90% Anglo-American.

Source of Information: Charles D. Mayer, Carrizozo, New Mexico.

The North and South Homestake Mines

by Edith L. Crawford

The North and South Homestake Mines are located on the east slope of Baxter Mountain in the White Oaks Mining District. The town of White Oaks, New Mexico, is located on the east side and at the foot of Baxter Mountain.

On the 14th day of August, 1879, three prospectors, John E. Wilson, Jack Winters and George Baxter, were prospecting on the east slope of Baxter Mountain. Being tired they sat down to rest and while resting John E. Wilson was idly digging with his prospector's pick when he broke off a piece of ledge and greatly to his surprise found the rock filled with gold. The three men at once staked out a claim and called it the "Homestake Lode." Soon after making the strike George Baxter sold out his interest in the claim to John E. Wilson and Jack Winters for a saddle horse and $43.00 in cash. He then left White Oaks and was never heard of again. The mountain was named "Baxter Mountain" for this same George Baxter.

After buying out the interest of George Baxter, Wilson and Winters divided the claim and Wilson took the south half and Winters took the north half. They called them the North and South Homestake Mines. The "Homestake Lode" was the first big gold strike made in the White Oaks Mining District.

Shortly after this division of the North Homestake Mines was sold to Jas. M. Sigafus for $50,000.00 and in twelve months the mine returned the purchase price and earned a profit of $10,000.00. It was stated that this mine yielded $35,000.00 from rich pockets in two days.

In the year 1882 the North Homestake shaft was down 40 ft. with drift in to top of shaft 60 ft. In 1892 the shaft is down 695 ft. with average stations every 40 ft. and an aggregate of about one mile in drifts. This is on the north end of the mine. On the south end, a shaft 310 ft. deep, with levels 50 to 153 ft. in length. The mine and belongings are owned by Jas. M. Sigafus in 1892.

The total yield of this mine up to January, 1904, was $525,000.00.

The South Homestake Mine was the first mine on Baxter Mountain on which large money was expended in developing. In 1882 its pay roll consisted of five men and the methods used were crude. At that time two drifts, each about 150 ft. existed, a shaft about 18 feet deep, and the discovery shaft on the line between the North and South Homestakes about 30 feet deep. In the year 1892 there are 1200 feet in shafts, 1500 feet of tunnels and over 2,000 feet in levels and drifts. In 1892 the South Homestake was owned by a company with improved machinery for working.

This mine has had some very hard experiences. In July, 1891, the Mine was discovered to be on fire, the fire having started the night before near the surface. The timbers in the shaft had burned down about sixty feet and the timbers burning off and falling set fire to the bottom. By hard work on the part of the miners the fire was extinguished after about twenty hours. Two men were trapped in the mine and lost their lives by suffocation. Their bodies were recovered after two days and were interred in the White cemetery.

The total yield of this mine up to January, 1904, was $600,000.000.

Jack Winters died in White Oaks, March 21, 1881 and was buried in the White Oaks cemetery. George Baxter left White Oaks in 1979 and was never heard of again. John W. Wilson died in White Oaks in 1892 and was buried in the White Oaks cemetery.

Source of Information: The information for this article was obtained from newspapers on file in the office of the County Clerk of Lincoln County, from the years 1889 to 1904.

The Old Abe Mine (I)

by Edith L. Crawford

Last November a year ago marked an era in the history of the White Oaks mining district, which has been the means of bringing it into prominence and has placed it in the front rank of the successful mining camps of New Mexico. This was occasioned by the discovery of a new ore chute in the "Old Abe" lead. Up to this time the camp, (though it had several valuable and paying mines and had taken an occasional spurt into life, only to die out again in a short time), was simply existing, living in the hope of a brighter future and a railroad to develop her varied and unequaled resources. Today these hopes are in a measure realized, for so long as the "Old Abe" continues to pour forth its riches, the prosperity of the camp is assured, while the prospects for the early construction of the railroad grow brighter every day.

In the month of November, 1879, J. M. Allen, O. D. Kelsey, and A. P. Livingston located a mining claim on Baxter Mountain which they designated as the "Old Abe Lode." Assessment work was done on this location until 1882. Some good ore was taken out but it appeared in deposits and they were unable to find the vein from which these deposits had their source. No work was done in 1882 or 1883, and the locators, believing there was nothing there, dropped it. In 1884, Messrs. John Y. Hewitt, Wm. Watson, and H. B. Fergusson located the White Oaks and Robert E. Lee Lodes, which together embraced a part of the ground of the Old Abe claims. These gentlemen expended considerable money running a tunnel to a distance of 100 feet. A small vein was found some time afterwards a little way from the mouth of the tunnel, on which a shaft was sunk to a depth of 70 feet. This last work was done in 1885, but was discontinued because the vein was so small that it did not pay to work it. In November of 1890, Mr. Watson went to work 152 feet south of the old shaft and opened on what afterward proved to be the same vein, and unearthed a bonanza. A rich body of ore was found a few inches from the surface, which has continued down to the present depth of 459 feet. The vein varies in thickness from four to six feet. Drift No. 2 was run along the vein and tapped the old shaft, thus securing thorough ventilation for a long time to come. Last December the old Glass mill was rented and operated for a time, but for the past three months the South Homestake twenty-stamp mill has been leased and is pounding out about 1200 tons per month. About 5,500 tons have been crushed since operations began.

The company does business under the name of "The Old Abe Co." and the mine is known as the "Old Abe," notwithstanding the real location names of White Oaks and Robert E. Lee. The fortunate owners have realized very handsome returns in gold bullion since their strike of last year, but how much is their own private affair. Several months since they refused an offer of $600,000.00 for the property. The wealth and splendor of King Solomon's mines fade into insignificance when compared to the wonderful riches of the "Old Abe," yet undeveloped. The owners and their interests are as follows: John Y. Hewitt...1/3; H. B Fergusson...1/3; Wm. Watson...11/48; W. M. Hoyle...5/48. Mr. Hoyle has charge of the mill, Mr. L. J. Banks superintends the mining and Mr. Frank J. Sager looks after the financial affairs. A sixty-horse power steam hoist is kept busy night and day.

At no distant day the South Homestake mine will again be in operation and it will then become necessary for the Old Abe people to erect a mill of their own. White Oaks is to be congratulated from the fact that all of the owners of this valuable property live here, excepting Mr. Fergusson whose home is in Albuquerque, and most of the money derived from the mines is expended and invested here. That many other "Old Abes" will yet be found here is as certain as death and taxes, and White Oaks will continue to grow and prosper until she stands foremost and the fairest city in this vast Territory.

Old Abe Strike

The Old Abe Company has felt satisfied that other veins, besides the one which they have been working so successfully for a year, existed within the boundaries of the Old Abe claim. To demonstrate this fact they have been driving a tunnel or cross cut west and north from their fourth level in the mine; commencing at a point twenty-five feet north of the mine shaft. Having drifted about fifty feet they came in contact with a body of ore six feet in thickness. This ore, like all other gold ore in this district, is free milling ore and is being crushed now by the Old Abe Company, and is running from $25 to $30 per ton. This ore has been followed, up to the present writing, for fifteen feet along the vein. This is an entirely new strike and runs northwest and southeast instead of in the same direction as the old lead upon which the company has heretofore been running. It cannot be positively determined yet that this is a true fissure vein but everything so far indicates that to be the case, and should it prove to be a true lead, it will favorably affect many of the other properties in Baxter Gulch. This strike is in keeping with the general belief of practical miners here, that White Oaks and other portions of Lincoln County, including the Jicarilla and Nogal districts, will be the greatest gold producers in the United States.

Sources of Information: Copied from the *Old Abe Eagle*, White Oaks, New Mexico. Dated December 18, 1891; The *Old Abe Eagle* was a newspaper published in White Oaks, New Mexico, by H. L. Ross, owner and editor.

The Old Abe Mine (II)

by Edith L. Crawford

Speaking with Colonel Prichard the other day about his first mining experience at White Oaks, he said: "I came here the first time in the fall of 1879. The Old Abe was located by A. P. Livingstone, the kindest hearted old man that I ever met. He was all heart and his habits as simple as those of a child. The old gentleman always claimed that the Old Abe was a rich prospect, and predicted that some day it would be a great producer. Poor old man; he did not live to see his prediction verified. None of us knew anything about the character of the ore deposits at that time and hence the mining we did was not always with the best of judgment. We did not dream of air shoots and chimneys then—these things have been learned in this camp by actual experience, after many thousands of dollars have been thrown away. Besides Livingstone and Allen, the Old Abe has been owned by H. J. Patterson, Chas. Buford and C. Ewing Patterson—Chase, Redman, and Speigelberg-Hockradle, and heirs of the original locators, Allen and Livingstone. Then Patterson Bros., Buford, Chase, Hewitt and Fergusson, and finally the present owners, Hewitt, Fergusson, Watson and Hoyle."

This mine has always borne a fine reputation on account of the rich placer gold found near and upon it. Various prospect holes were sunk, but not until 1890 was anything found to create very active interest. During the month of December Chas. Hamilton discovered croppings of value and he, Watson and Lund began a shaft which the owners soon took control of. For twelve months, ending Oct. 1, 1892, this shaft has produced 50 tons of ore per day, which has averaged between $15.00 and $20.00 a ton.

(Copied from *The Outlook*, dated April 24, 1908. *The Outlook* is a newspaper edited and published in Carrizozo, New Mexico by Carrizozo Publishing Co.)

No mine in White Oaks district is more intimately connected with, or interwoven in the fabric of the history of the camp, than the "Old Abe." The Old Abe vein is a true fissure varying in width from 3 inches to 16 inches, but attains the extreme width of 22 feet in the chamber known as the fish pond, between the 7th and 8th levels. A total depth of 1,375 (now 1,420 feet) is reached which is the deepest mine in New Mexico. Practically no water has been encountered and it is said to be the deepest dry mine in the world. The total output to January 1, 1904, is $875,000.00 in gold; but a little more than a trace of silver is present in the ore. Virgin gold in gypsum is one of the remarkable occurrences in the Old Abe.

The "Old Abe" is really a phenomenal mine. (John Y. Hewitt in *The Outlook*.)

Source of Information: *The Lincoln County Leader*, White Oaks, New Mexico Dated Oct. 22, 1892.

American Guide   Kenneth Fordyce
                                                        March 5, 1937

Tin Pan Canyon, Colfax County

This cayon got its name just before Blossburg came into being in 1882. The canyon extends westward from Dillon Canyon, where Blossburg lay, and is a few miles beyond the site of the deserted camp. Tin Pan Canyon got its name from a shining tin pan that a miner nailed to a post at the entrance of the canyon. He was going in ahead of some friends and he put the pan up to guide them to him. The canyon soon became known to everyone as Tin Pan Canyon. Information by Babe Bartolino.

(FACSIMILE) Tin Pan Canyon, Colfax County, Kenneth Fordyce, March 5, 1937, NMFWP, WPA #190a, NMSRCA

Coal in Colfax County-New Mexico

by Kenneth Fordyce

The presence of an immense coal field in Colfax county, New Mexico was a matter of common knowledge since the early '60s. Coal for domestic use was mined near Cimarron before the coming of the railroad in 1879, which heralded the beginning of an industry destined to become the county's most valuable economic asset.

The first mine opened in the vicinity of Raton was the Raton vein on the east side of Railroad canyon, opposite Lynn, New Mexico at the south entrance to the Raton tunnel (on the state line). A tipple was erected and the engines coming over the switchback were supplied with fuel from this mine until the opening of the Blossburg mines two years later.

Small mines, for supplying Raton trade, were opened in Climax canyon and the next canyon north of Climax. These mines were west of the present site of Raton, one on either side of the present Scenic highway; they were worked extensively at that time.

The first mine opened in Blossburg canyon was located on the east side of the canyon near the present site of Swastika (a present day coal town). The next mine was in Seeley canyon west of the present site of Gardiner (a present day coal camp). The third was in Dutchman's canyon, which was west of the site occupied by Blossburg (an abandoned coal town). A subsidiary company of the Santa Fe railroad, known as the Raton Coal and Coking Company, operated these mines until 1891 when Mr. E. P. Ripley, the incoming president, inaugurated his policy of abolishing all side issues and returned the operation of these mines to the Maxwell Land Grant Company.

Following the sympathetic coal miners' strike in 1893, the Blossburg mines were closed for several months.

The St. Louis, Rocky Mountain & Pacific Company was organized on July 1, 1905. This company took over Blossburg and the organization operated both Blossburg and Van Houten. After a short period of further activity Blossburg was finally abandoned and practically all evidences of the existence of a town of three or four thousand people have been obliterated. A half dozen houses and a few foundations mark the spot of the former flourishing coal town. Van Houten was originally called Willow, but was later named Van Houten, after the President of the coal company. Van Houten, or Willow, in Willow canyon, opened in 1903.

From Preston and Koehler Junction, the road and the railroad lead up Prairie Cow Creek to another coal town which was started in the spring of 1906. They called it Koehler; it has not been used for a number of years and is practically uninhabited.

At Gardiner, another coal camp of the company, coking ovens were built between 1898 and 1906.

Brilliant, another camp up Dillon Canyon just above Tin Pan canyon, was started early in 1906. This camp was shut down after twenty years of operation.

Sugarite, in the Sugarite canyon east of Raton, was opened in 1912.

The company camps that are working today are: Gardiner, Swastika, Van Houten, and Sugarite.

The Santa Fe, Raton & Eastern railroad built eastward from Raton to Yankee, which was a live coal mining town in the early 1900s, operated by another company. It is now practically abandoned.

The Phelps-Dodge Corporation has operated a coal town at Dawson, New Mexico, thirty five miles south of Raton for a great number of years. At one time this town had a population of some 5,000 or 6,000 people. Its population is much less than that today but it is still a large camp, and operating. Dawson in located on the Union Pacific railroad.

Many independent owners operate mines in Northern New Mexico, for coal is abundant and it adds greatly to the amount of business done in Raton each year.

Source of Information: Jay Conway, Raton, and (Observation).

Elizabethtown

by Kenneth Fordyce

They call it E'town today. A trip through the northern end of the Moreno Valley will show that the town has decreased in size even more than the name has been shortened. The busy little city of Elizabethtown once lay in the northern end of Moreno Valley. The few houses which comprise E'town today are on state road number 38 and lie a few miles north of Eagle Nest, which is on US 64. State road 38 is the one which leads on into the Red River Country.

The discovery of gold in New Mexico is of ancient date. The first discovery of gold on the western slope of old Baldy Mountain in Colfax County can be attributed to an Indian. For years the Ute and Apache Indians roamed over the Mountains of Colfax County in quest of game. They frequently came to Cimarron where they received their supplies from the government agency. On a trading expedition to Ft. Union an Indian exhibited some copper float. This was in the early '60's and copper was in great demand. A group of friends paid the Indian a small amount of money to show them where they had found the copper. In 1866, this group located the "Copper Mine."

The group from Ft. Union discovered gold on Willow Creek in large enough quantities to cause great excitement. As it was late in the season, and the winters severe in those mountains, they swore each other to secrecy, and returned to Ft. Union to wait for the spring when they could begin gold mining operations in earnest. Long before the spring of '67, the news of the rich diggings leaked out and spread over New Mexico and Colorado. Crowds started pouring into the Valley. The original partners staked a plot on Willow Creek, and other groups staked out all available land along the creek bed. About this time gold was discovered at Elizabethtown.

The entire north end of Moreno Valley was found to be sprinkled with gold which lay close to the surface. This called for placer mining which required great quantities of water.

The great number of people arriving daily started thoughts and discussions about the building of a new town. A plat and survey were made. It was decided to call the new town Elizabeth after Elizabeth Moore, who was later Elizabeth Moore Lowrey. Elizabethtown grew rapidly after its birth in '67 and became a place of great importance, as was evidenced by the fact that soon after when Colfax County was taken off of Mora County, Elizabethtown was the first county seat. It had a mayor and a full set of city officers and was the first incorporated city in the territory.

Water was the great need now to successfully wash out the gold which they knew was there. The few streams in the valley failed to supply enough. A competent engineer was employed. He found a good supply of water in Red River, about ten miles west of the mines, and planned a way to bring it to the valley.

A company was organized. The line of "the big ditch" was surveyed and work was commenced. The ditch was completed before the end of the year 1868. The main ditch was forty-two miles long and cost $280,000.00. Branch ditches, about eight miles long, and three lakes high up in the Red River Mountains, were built at the additional cost of $20,000.00. On account of the seepage and the evaporation in coming such a long distance, the company could not deliver enough water and found an unprofitable enterprise on its hands.

In 1875, the ditch was repaired by a new owner and was made to carry a full head of water. Four little giants and one hydraulic were all that they could run with their water supply, but the ground was exceedingly rich and the output very large. From 1867 until the gravel surface had been worked, and the supply of gold had diminished to a point where further operations were unprofitable, millions of dollars worth of gold went through the little town of Elizabethtown on its way to the markets of the worlds.

Like many another boom-town, Elizabethtown is today a ghost town. The buildings have been removed or destroyed. A few scattered buildings, some occupied and some in ruins, an occasional piece of machinery, rusted and half buried, and memories of a gay town where life was fast and gay, are all that remain.

The abandoned spot seems more realistic when you search out some old-timer who lived at E'town during its brief existence and he tells you some of the many hair-raising stories of shootings, neck-tie parties, and stick-ups by the wild desperadoes of that day. Then there are the tales about great quantities of gold which were never dug out, and it makes you wonder . . . oh well, that's E'town.

Source of Information: Facts obtained at Raton Public Library from an advertising brochure printed in about 1901.

Description of a Mine: Santa Rita Copper Mine

by Mrs. Mildred Jordan

The Santa Rita District is in Eastern Grant County. The town of Santa Rita is near the center of the district and about twelve miles East-Northeast of Silver City, the county seat. It is the terminus of a branch line of the Atchison, Topeka, and Santa Fe railway, leaving the Deming-Silver City branch at Whitewater.

The only copper mining of importance in the United States antedating operations of Santa Rita was the native copper deposits of Northern Michigan. Mining at Santa Rita by the Spaniards began in 1801, the copper being shipped to Chihuahua and Mexico City under contract to supply metal for Mexican coinage; production during this period is said to have amounted to 4,000,000 pounds of copper annually.

It has been estimated that production up to 1904 amounted to 80,000,000 pounds. From 1904 to 1909, when the Santa Rita mines were purchased by the Chino Copper Company, annual production averaged 3,500,000 pounds of copper. Production dropped during the next two years, presumably because of preparations for steam shovel operations, which were started in 1910. The first ore was milled a year later. From 1911 to 1930 inclusive, representing the period of steam shovel operation, production amounted to about 1,040,000,000 pounds of copper, valued at about $180,000,000.

The ore reserves of the Chino mines amount to more than 100,000,000 tons averaging about 1.27% copper, enough for about thirty five years at the average rate of production prevailing under normal conditions. Prospecting for new ore is by means of churn drills generally to a depth of nine hundred feet. The mines are now owned by the Nevada Consolidated Copper Co. Ore is treated at the company mill at Hurley, ten miles south of Santa Rita, to which it was hauled with standard railroad equipment. The mill has a capacity of 13,500 tons a day. The concentrate was shipped to the American Smelting and Refining Company at El Paso, Texas. Santa Rita lies in a well defined basin formed by a local widening of Santa Rita valley. The rim of the basin is highest at the Santa Rita mountain, which has an elevation of 7,365 feet, about 1,600 feet above Santa Rita Creek. The mountain rises along a steep erosion escarpment 400 to 1200 feet high and is cut by many steep gulches and canyons, a continuation of the escarpment from the Eastern rim of the basin. The remainder of the basin rim was originally formed by a series of hills rising 100 to 450 feet above the basin floor, but some of the hills have been removed, wholly or in part, by steam shovels.

The intrusive rocks at Santa Rita include sills, stocks, and dikes. Extrusive erosions followed the period of intrusive activity and uncovered the intrusive rocks in several places. The ore of the Santa Rita deposits is somewhat richer than the disseminated ore in several other similar districts in the southwest. As in all deposits of this class, the line between ore and waste is an economic one, controlled by the price of the metal and the cost of production.

Pike, in 1807, refers to a copper mine west of the Rio Grande, yielding 20,000 mule loads of metal annually; this is thought to be the Santa Rita mine. This mine was discovered in 1800 by Lieutenant Colonel Carrasco. Through the aid of an Indian, Twichell records the speech of an Indian chief when a negro went to the Pueblo of Hawaikuh.

The first American to visit the mine was James Pattie, a trapper, hunter and explorer who, in his "Narrative," tells of the workings of the mine by a Spanish Superintendent, Juan Onizz, for the Spanish owner, Francisco Pablo Legara. Pattie and his associates finally leased the mine for five years, agreeing to pay $1,000 a year, and apparently worked until 1827, when Legara was exiled as a Spaniard. The implication is that the mine was abandoned. The Indians at that time occasioned a great deal of trouble. The Patties in 1825 made a treaty of peace with the Apaches. One day the young Patties (it is told) were hunting deer when they discovered the trail of Indians approaching the mines. Following, they came upon the Indians, who immediately fled. They were captured and questioned and one of them told to leave the camp and tell his chief to come with all his warriors and make peace. On the fifth day of August, eighty Indians appeared and told the Patties that they had no objections to a peace with the Americans but would never make one with the Spaniards. When asked their reason they replied that they had been at war with the Spaniards for many years and that a great many murders had been committed by both sides.

A dishonest agent, who Pattie entrusted with $30,000 in gold to buy the mine, promptly left the country and was never heard of again. Patties were ruined men.

Robert McKnight possessed the mine in 1834. In 1862 General Sibley owned the mine. 1873 M. D. Hayes bought the claims. 1880 J. Parker Whitmey purchased it. 1897 it was owned by the Hearst Estate. The mill was started at Hurley in 1914. Many traces of Spaniards' works have been discovered in the mine district.

Source of Information: Mrs. J. J. Burr—Newspaper article, Silver City, New Mexico.

Description of Point of Interest: Steeple Rock Peak

by Mrs. Mildred Jordan

Steeple Rock Peak: is located in the South West part of Grant Co., Carlysle, an old mine and mining camp located in the peak. In early days, such mining was done, some is still being done, by different men, who have leased places to work on. The ore findings were gold, silver and lead, mostly gold. Also in the early days, there were killings done by the Indians, in places where the old trails used to be. There are now many ranches, cattle raising is done in the district, the timber is very scarce.

Water runs in the Gulch most of the time; very little stock is raised. Some game is found, such as deer, bear and lions. The Carlysle Mine was once the richest gold mine in the country, but hasn't produced very greatly since the nineties.

Being sold by Mr. Johnson, Green Hale, and Frank Weldon to an Oklahoma Company, the mine is again being worked. The mine has already produced over $3,000,000 worth of gold. Recently Mr. George Utter of Long Beach, California purchased it. The mine and shafts are so run down and in need of new material, it will take much money to get it in running condition to run again in a big way. There is farming on the river south of Steeple Rock.

At Verdon, a voting place near there, about four hundred people registered this election. There is much beauty and scenery down the Cliff road. Herbert Hoover first came to Steeple Rock Peak to look over the mine after he graduated.

Source of Information: Judge B. B. Ownby, Lordsburg, New Mexico.

Sad Disaster at Mogollon

by John Looney

Taken from the *Santa Fe New Mexican*, Aug. 20, 1896. "Prosperous Mining Camp in Socorro County Visited by a Destructive Cloud Burst." Two Bodies Found—Twenty Missing.

Colonial, Deep Down and Helen Mining Companies are Sufferers to Amount of Thousands of Dollars. New Machinery Ruined.

Aug. 20–A special to the Times from Mogollon, New Mexico received today says: A terrible cloud burst occurred about 4 o'clock yesterday afternoon.

John Knight, a miner, of Georgetown, was drowned.

Twenty others are reported missing, but so far only two bodies, Thomas Knight and an unknown Mexican, have been recovered. About 100 families have been rendered homeless. Thirty houses were washed away.

The property of the Colonial Mining Company, of Boston, Mass. has suffered to a large extent, the assay office, mill house, powder house and blacksmith shop being washed away.

It is feared that the Mine is filled with mud and water.

The manager and assayer had a narrow escape, being assisted to the bank by means of ropes.

The Deep Down Mining Company, of Kansas City, lost its main office and assay office.

This place is situated in a deep canyon between high mountains. The water in the streets was eight feet deep. The storm was general in this section of the territory.

Advices from Graham state that about 4 o'clock yesterday a cloud-burst on the mountainside caused a flood at the Confidence Mine belonging to the Helen Mining Company of Denver. The flood carried away the shop and supplies of the mine.

The mine horses loaded with ore for the mill were washed over a steep precipice and killed. The men who were working at the mouth of the tunnel barely escaped with their lives.

It is feared that great damage was done on the other side of the Mogollon district. Nothing definite can be learned on account of the telephone communication being broken.

There was at least twenty persons, mostly miners, living right in the track of the great wall of water. They occupied, for the most part, adobe dwellings. These have been washed away and the occupants have not been heard from. Some may have been warned in time to get out of the way.

The loss to the mining Companies will amount to thousands of dollars. Expensive machinery has lately been put in and much of this is wrecked.

Taken from *The Santa Fe New Mexican*, Aug. 20, 1896.

The wagon road down Silver Creek from the town of Mogollon has been almost entirely destroyed by the flood. A new road from Whitewater will shorten the distance about nine miles between Silver City and Mogollon and provide a much better and easier route than the former road. The friends of the Socorro road are now agitating for the new road in that direction, which would cut off the trade from Silver City and divert it to Socorro and other towns. The merchants and business men of Silver City should call a meeting and raise a subscription to aid in the immediate construction of the new road. Now is the time when the Mogollon trade will either be saved or lost to Silver City. Our business men should act promptly. – *Silver City Enterprise.*

Citizenship Papers: Celestino Carrillo

by Ernest Prescott Morey

It was in the month of June, in the year of 1863 that many peons in Old Mexico were having a gay wedding party in the small mountain town of Dolores, in the state of Durango. Celestino Carrillo had promised before God, to love and cherish Teodillia Garcia. For several days, the people of this small town in the mountains were enjoying the celebration. The bride and groom were fifteen years of age.

Several mornings after the festivities, Celestino and Teodillia, having bid good-bye to the wife's parents and relatives, were traveling slowly down the steep mountain trail, with three burros, on their way to Chihuahua City. Arriving in Chihuahua City, Celestino took his bride to the abode of his father, where they were given a royal welcome, and another dance, or baile. For three years Celestino made a living for himself and his wife by loading and unloading the freight caravans, which traveled over the old Rio Grande Trail to El Paso and Santa Fe.

Many the story of adventure and wealth Celestino heard from the long haulers about this marvelous northern country. Each evening when the folks of Celestino were assembled around the table for the meal, he would tell them what he had heard during the day and his eyes would glow at his own stories.

His father would say, "Hijo (Son), you have heard the rosy side of things only. That trail is rugged and long, and not only filled with the dangers of nature, but it is attacked by Apaches and bandits."

Celestino told me that his eyes would fill with tears, because his hopes of ever going to the north country would be dimmed by his father's words.

He would say to his father, "You traveled that trail until you hurt your back."

"Yes, but you're only a muchacho (boy)," his father would answer.

"Well, I've got to make a living and I only make a few centavos here a day, while I could be making plenty if I were in that north country."

After more warnings, the old folks would retire. Celestino and his wife would leave the table and go to the placita, where they would sit among the pepper trees and talk about that mysterious Santa Fe where the great Archbishop, the friend of his father and mother, had his abode.

One evening, shortly after this conversation, Celestino dashed into the house, breathless from running, and his eyes sparkling like the dew drops vanishing in the early morning sun. "Padre! (Father!) Madre! (Mother!)" he shouted. "Adolpho Hernandes has given me a chance to take my wife and go to El Paso, Texas. He will give me a dollar a day, and he says that there are great places for ranches in the northern country. He says that a young man of courage can make a fortune there."

All the pleadings of his Padre and Madre could not dissuade the young man from making the trip. Two days later, Celestino and Teodillia were in one of Hernandes's big teams, traveling northward. There was a guard of forty Mexican soldiers going with the train to Santa Rita, New Mexico. This train had been twenty days from Chihuahua, when a band of Apache Indians attacked it early in the morning at a place called Vicitas Canyon. For two days, the soldiers and men fought the Indians, finally driving them off. By the grace of God, there were no casualties among the soldiers or the men of the train.

Young Celestino arose in the estimation of the soldiers and the travelers of the caravan, because of the way he performed during the attack. Even little Teodillia's eyes glowed as she would hear the praises that were showered upon her husband.

Three weeks afterwards, the wagon train reached El Paso. Celestino was paid forty-one dollars in American money, and then went off to secure quarters for himself and his wife, promising the commander of the train that he would be ready to continue his journey to Santa Fe and the Archbishop.

Yet, fate had to play her hand and change the course of this young couple. Seated that evening in a small restaurant, the commander of the soldiers came to their table while they were eating. He sat down with the young couple. This was in August, the year of 1866.

"Celestino," said Medina, who was the commander of the soldiers, "why don't you and your wife go along with us? We're going to Santa Rita. I can secure you work there for fifty cents a day."

The next day, Celestino and his young wife set out for Santa Rita with the soldiers. After bidding the commander of the wagon train goodbye and telling him of his change of plans, he joined Medina.

Toward the close of September of the year 1866, Celestino, his wife and the soldiers arrived in Santa Rita. Medina secured him a position at the old Spanish fort. For three years, Celestino worked at the fort. Then he was given an opportunity to work in the copper mines at the same rate of pay, fifty cents a day. The couple lived in Santa Rita for thirty-six years. They managed to live very well, but did not accumulate any money in all this time. Teodillia presented her husband with two children, Clair and Sofa. Young Celestino had made many trips around the surrounding country during this thirty-six years, and he had longed to own a ranch in the Mimbres Valley, which he considered a marvelous country.

In the year of 1912, while prospecting around Santa Rita and Northeast of the copper mines, he found his pot of gold at the end of the rainbow. He struck a gold vein, and staked himself a claim, which he recorded in due form. Then he made application for his citizenship papers, which he had neglected for so long, because he knew that he would lose the mine if he did not do other-wise. He took sack after sack of gold from this claim and hauled it to Silver City, where he exchanged it for money. With this money he bought himself a lot of the surrounding property and built himself a house right over his diggings. In due time, he was granted his citizenship papers.

Today he owns thirty-five or forty houses and a store in Santa Rita. He has also bought himself a large ranch in the Mimbres Valley.

The Chino Copper Company have offered him seventy-five thousand dollars, time and again, but he has refused all of these offers.

Now in the twilight of his life, he and his wife sit in his garden among the fruit trees, watching the mountains which surround them. Five other children have been added to their family, and like eagles who have left their nests, they have gone their various ways. The old people of that abode, in the Mimbres Valley, often think back to their childhood days and their long trip from their old country to the new.

As I, the writer, sit on an old log in the back yard of an old rancho just across the way, with an old lady who was born and reared in this great country, I am listening to her story.

"Celestino," she says, "is skeptical of all men. He is taking his ease as he sits in his garden dreaming of the past."

The old steam shovel in the Santa Rita Copper Mines pit is digging, digging, digging toward the property and gold claim of this boy of yesterday, who is the old man of today in the Mimbres Valley.

Maybe the old grim reaper will carry him away before the shovel reaches his El Dorado, but if not, he may find himself fighting a greater fight for his gold than he did for his life on that trip from Chihuahua.

Source of Information: As told by Jessie Swartzs, who is 50 years of age. She was born and reared on the Mimbres. She is the daughter of Jim Swartzs, who came from Germany seventy years ago. Her birthdate was August 30, 1888. Her address is Mimbres Valley, New Mexico. She can also receive mail in Silver City, New Mexico, 106 Spring Street.

Bledsoe, T. F.   6/10/37   cl- 300

GLOSSARY OF MINING IDIOMS USED IN NEW MEXICO

-B-

blow-out — An immense outcropping of ore more bold than a vein.

-C-

chimney — A shoot of ore.

claim jumper — One who steals another's claim.

-D-

diggings — Mines.

dust — Placer gold.

-F-

float — Gold or silver bearing quartz or galena scattered from lodes.

-G-

Gambucinos — Petty miners who work independently.

-H-

high-grader — One who steals high-grade ore.

hill nutty — A prospector who has stayed in the hills too long.

-J-

jump a claim — To steal another's claim, or property.

-P-

pay dirt — Rich ore, generally used in connection with placer mines.

pinched out — A vein that disappears or runs out.

placer — Mineral deposits, not veins in place; generally means free gold in dirt or gravel.

pockets — A small deposit of gold or other ore

—R—

red beds   A Permian formation.

rich strike   A rich silver or gold deposit.

—S—

salting a mine   Placing foreign ore in a mine so
                        that it appears natural.

source   The mother lode

strike it rich   To discover a good deposit of
                        mineral; made good.

—T—

tantes (a)   Rawhide buckets used for conveying
                        waste and ore to the main shaft,
                        and carried on backs of Indians.

to make a stake   To make a fortune.

trass   Volcanic earth found in certain
                        localities of New Mexico.

—V—

volcanic waters   Thermal springs.

Reminiscence of an Old Prospector

by Ernest Prescott Morey

Santa Rita, situated in the northeastern part of Grant County in the state of New Mexico is one of the oldest mining districts in the west. Its date of operation is said to reach back to the time that the white man's foot had not touched the soil of the Americas. The Indian had made use of the red metal long before the coming of the Mexican, or Spaniard. Many evidences of this fact remain as proofs, such as copper arrowheads (that when shot at a close range produced a deadly wound), metal jewels, and ornaments of all kinds. An old Spaniard by the name of Pasquel Varges, in Silver City, New Mexico says that the Indians of some race that lived around in the Mimbres even used copper for domestic utensils. (However, not being able to get authentic proof of this fact, I do not write it as a certain statement.)

The country around Santa Rita is composed of flat rolling hills, which continue on to the west. The mining district is bounded at the north and south by rough escarpments of rhyolitic lavas; on the east by the serrated ledge of limestone mesa that extends on north to the Mimbres Valley. The principal deposits of copper found at Santa Rita lie in an area of about three fourths of a mile square. This area is known as the Santa Rita basin.

The famous copper mines and this basin were discovered in the year of 1800 by the Spaniards. At the time of its discovery, it is said that the copper was so abundant that it lay bare, and rich at the foot of "The Kneeling Nun," an old legendary landmark.

Much mining was done here by both the Indian and the Spaniard. Mining had its beginning in the year of 1804. At this time, the workings were known as the Romero workings. In the latter part of the eighteenth century, Colonel Carrasso of the Spanish army was stationed in the Santa Rita area. He was a very good religious man. It was said that he tried to be fair with the Indians as well as the people of his own race. An old prospector said that he was a good, as well as a handsome man. He said Carrasso was tall, well built, broad shouldered, and wore a mustache and sideburns. He had left a wonderful home in Spain, where he had been reared on a large estate, or rancho, which was owned by his father, a Don.

One morning, an Apache Indian rushed into the fort at Santa Rita having been bitten by a rattlesnake. Carrasso, who at the time was drilling his detachment in the cool of the morning, hastened to the Indian's side, and immediately treated him for snake bite. The Indian remained at the fort for several weeks. During this time, he was treated so well that he learned to like the Spaniards. He knew that they had been his friends. Upon the day that the Indian was able to leave the fort and go back to his tribe, he called Colonel Carrasso aside to show his gratitude and bid him good-bye. It is a common story around Santa Rita that the Indian said, "Me show you the red metal, you follow me."

Carrasso became animated and left the fort alone with the Indian, for here, after many years of searching by the Spanish expeditions, he was going to learn the secret of where the Indians found the copper that had made their arrowheads and ornaments.

The Indian, true to his word, showed Carrasso where the copper was. It is contended today that Carrasso was the true discoverer of the Santa Rita copper mines. It is said that after that Indian showed Carrasso this wonderful mine of riches, he walked away toward the north to the Mimbres Valley. This was the last that Carrasso or the Spaniards ever saw or heard of him. Don Carrasso, being a very sentimental man, watched the Indian until he had gotten out of sight, across the limestone mesa.

Arriving back to the fort, he immediately sent off for a great friend of his, who was a wealthy merchant and banker in Chihuahua, Mexico. Manuel Elgua, the friend, left Chihuahua City at once and arrived at the fort, in Santa Rita, within a month's time.

The Colonel and Elgua then made a dicker to furnish copper to the Mexican government for her coinage. The two men received in Mexican money, what would have been equal to about sixty-five cents in American money at that time, for each pound of copper sold. It was a very prosperous business, because it is known beyond a doubt that twenty thousand mule loads of copper were sent into Mexico annually.

Many are the stories of how this pack train was attacked on the long journey to Mexico. Wars and battles with the Indians were continuous. It has been said that when an Indian was captured, he was made to work as a slave in the mine for his punishment. Some of the Indians died while working in the mines. Workers were scarce, so Colonel Carrasso succeeded in having convicts sent in from Mexico to work the diggings.

In the year of 1815, or perhaps sooner, Carrasso sold his interest in the Santa Rita copper mines to Elgua. Carrasso did not understand, or realize the wealth of this vast holding, and neither was he a mining man. Elgua continued working the mine until the American occupation. The business proved so profitable to him that he became exceptionally rich within a short time. He used every means to protect himself and his men. He had a large adobe fort erected close by the workings, in order to protect his miners from the Indian raids.

In the year of 1831 a flourishing trade between California and New Mexico sprang up. A pack train of mules were sent over the Gila Trail, laden with copper from the Santa Rita mines, as well as New Mexican blankets and other goods, which were exchanged for mules and horses.

It was about this time that John B. Magruder came to Santa Rita. In his command was a young man known as George Edwards, who was thirty-five years old, having a beard and a Van Dike, and he was a native of Georgia. About the same time that John B. Magruder came to Santa Rita, Colonel Sumner, who was in Santa Fe, established a fort in Santa Rita for the protection of the American interests and trade there. This fort is known as Fort Webster, and its remains are still to be seen.

Magruder and Edwards, who were in charge of this fort, had their hands full trying to defend the miners, etc., from the Apache Indians. They were the bitter enemies of the Americans, just like they had been the enemies of the Spaniards. There were two noted chiefs in the district, who were blood-thirsty and cruel. The first and foremost was known as Mano Colorado (Red Hand). The other was Cochise.

One evening Edwards heard a commotion in the horse corral,

while he was on guard. Upon investigation he found that the Apaches had stolen several head of fine horses. Reporting to his commanding officer, he was told to pursue the Indians. This was just what the Indians wanted. With twenty men Edwards rode about fifteen miles from the fort, following the trail the horses had made. He found that the trail branched off, and the marks of the stolen horses could be plainly seen, heading off in a northwesterly direction to the Mimbres Valley and the Black Range, while the other went northeast.

Edwards and his men pulled up for a short consultation. Like all young men who are headstrong, Edwards ordered his men to continue the pursuit along the trail headed for Mimbres. Old Joe Louis, who had served under General Wood and General Taylor at Chihuahua and Monterey, said, "Edwards, I don't want to try to tell you anything, but I've spent a number of years fighting these pesky redskins, and I am afraid they are up to something. I believe they are setting a trap for us. They are traveling much too slow to suit me. I believe it would be better for us to turn back."

Edwards told Louis that he was in command of the detachment, and that Louis would do as he said.

Louis did not answer him, but instead dashed off with the detachment as directed, in the wake of the stolen horses. The search continued all night, but at dawn the detachment was surprised by the wild yells and screams of the red men in the rear.

Edwards turned about and found to his dismay that he truly was caught in a trap. It was a well laid trap, he admitted at last to Louis, who was riding by his side. He gave orders to his men to seek shelter behind a large rock and some stunted piñon trees, and looked over their predicament. "There must be three hundred Indians," said Edwards.

"Yes," said Louis, "We are greatly outnumbered, and I believe there is only one thing for us to do. Someone should try to reach the fort, and bring help."

Edwards agreed with Louis this time, but he knew that it was impossible to send someone to the fort at that very moment. Every man was behind their shelter fighting for his life and several of them had been wounded seriously, already. Again and again on that mesa Indian after Indian fell from his horse. After three hours of terrific fighting, the Indians were repulsed and had drawn off to considerable distance. During this lull, Edwards called for volunteers to take a chance to ride to the fort. Sidney Butler and Joe Louis both tried to get the chance to go, but Edwards chose Butler. He told Louis that he needed him with the detachment.

Butler reached a horse and headed for the fort. He had gained considerable distance, when the bloodthirsty screams rent the air. The Indians knew what he was doing. Several young bucks mounted their horses and set out in hot pursuit. Butler rode swiftly on, not looking back. He had almost reached some small scrub timber, when an Indian shot his horse. To the sorrow of his comrades, he was brought back to the main body, where they immediately proceeded to make him run the gauntlet, after which, they burned him at the stake, before the very eyes of his comrades.

This filled most of the soldiers with horror, but old Joe Louis looking out over the top of a large rock, said, "All Indian fighters know that they are going to get just that. Those very Indians are going to pay for their plundering and murdering."

Some soldier shouted that it was going to rain. Everyone hearing this looked up into the heavens, watching the smoke that burned their comrade's body. Then it dawned on Edwards that Butler's fate would come to the rest of them if they did not take advantage of the lull while the savages were burning Butler's body.

Then he said to Louis, "Don't you think we all ought to make a dash for the fort?"

Louis told him that he was positive that they should, there was no other chance.

Orders were given and the stolen horses were forgotten. The men were to all make a desperate fight for their lives and whoever got to the fort was to bring help to the rest. No sooner had the orders been given than all the men were under way. The Indians were pursuing them in a terrible chase. What happened to most of the soldiers, no one knows.

Several hours afterwards, the lookout on the wall saw a jaded horse and rider coming toward the fort. It was old Joe Louis. He dismounted and told his story to Magruder. Immediately the bugle sounded and a hundred and fifty horsemen left the fort to seek their lost comrades. No Indians were to be seen as they rode along, but just before dark, Edwards and one other man were found. They were both afoot, their clothing torn almost to ribbons, and their feet badly battered and bruised from the climbing and stumbling over the rough country.

The next day, Edwards with almost a hundred men went out searching the surrounding country looking for traces of his lost comrades. On this search, he had the great fortune to come face to face with the dead body of the terrible chief, Cochise. He was dead from a gunshot wound. Close by his body, was the body of one of Edward's soldiers with his gun tightly clutched in his hand. Not a great distance away from this scene, a ghastly sight met their eyes. Another soldier had been killed and his body thrown into an ant hill. His body had been so badly mutilated that none of his fellow soldiers could recognize him. After several days of searching the group of soldiers returned to the fort without finding any trace of the rest of the troop, or sighting one live Indian.

Some time after that, Lieutenant John B. Magruder caught Red Hand and put him in the guard house for the night. In the wee hours of the morning, the American soldiers murdered him in his cell.

This is only one of the many stories that go to show what a terrible price the Santa Rita copper mines have cost the Americans.

This famous old mine at Santa Rita is still in action. Much work is still being done there. Two other important mines are located in Tyrone and Fierro, in this county.

In Hurley, not far from Santa Rita, there is a copper mill in which the copper is separated from its ore.

Silver City is the business center for this copper region. Santa Rita boasts that they possess the largest open pit copper mine in the United States, if not in the world. The ore is mined almost wholly by the steam shovels.

No one knows just how much copper was, or has, been mined from this pit. Lieutenant Pike, in 1807, estimated that the mine was producing twenty thousand mule loads annually for the Mexicans, or Spaniards.

Many of the old prospectors say that even as near back as thirty years ago, that the ore could be found bare on the foothills of Santa Rita.

The town of Santa Rita is supposed to mark the very spot where copper was first discovered. At the time the town was first founded, copper was the most important metal being mined in the state. At present, it ranks second, since the demand for zinc is foremost.

Source of Information: Celestino Carrillo (Old Prospector), age 80, Mimbres Valley, New Mexico; Otho Allen, age 53; John Wessley Allen, age 87, Silver City, New Mexico Present address of John Wessley Allen, Haber Springs, Arkansas.

Tyrone Turquoise Mines

by Ernest Prescott Morey

Years before the Spaniards came to New Mexico, the Pueblo Indians were working Turquoise Mines in a place that is now called Los Cerrillos, and is close to Santa Fe. These mines are the most important Turquoise Mines in the world and have been worked to some extent by various people ever since the coming of the Spaniard.

The Indians, with rude stone sledges, without the aid of iron, steel or explosives, broke away huge masses of rock in search of this fascinating ornament, which to their minds possessed some vague supernatural power.

In the southern part of New Mexico and in Grant County there is a town called Tyrone and this town is situated on top of a Turquoise Mine. This mine has never been worked, since it was rendered useless by the little village that was erected over it. Very slowly, but surely, this mine is turning to copper. Three miles above Tyrone, up in the mountains are another set of Turquoise Mines. They are not noted like the famous mines that are located near Santa Fe, but they have at various times produced fortunes for their owners. No one seems to know the exact date of the discovery of these mines, but in the year 1888, a man known by the name of Tom Parker became the owner of them.

In the year of 1903, there were three men who came to the United States, after hearing about the amount of Turquoise that could be produced from these mines in the mountain above the town of Tyrone. Very little was known about these men, except that one, Felix Vogel, was from Germany, another Louis Rothchild, from England, who belonged to the famous Rothchild family of England, and the other, Adolph Armmene, from Paris, France. How these three men of such different nationalities met and later bought Tom Parker's interest in the large Turquoise Mines is unknown. They gave Tom Parker a large sum of money for his interest in the mine. He appeared to be well fixed for the rest of his life.

Adolph Armmene was in business with Tiffany and Company, the big Jewelers of New York. Many valuable Turquoise sales were made to this company while these three men worked these mines.

In the actual working, and improvement of the mines, Louis Rothchild invested more money than the other two men put together. In the first year or two of his working, he invested thousands of dollars. The managing and working of the mines was almost solely done by Louis Rothchild and Felix Vogel, since Adolph Armmene spent most of his time attending to business in New York. Felix Vogel was made manager of the mines, and was often seen in them working with his men.

These three men made a very prosperous living from their Turquoise sales. Finally, Felix Vogel, after accumulating a good-size fortune, returned to Hamburg, Germany. There he married Frauline (Miss) Corinne Helman. They returned to this country and to Tyrone, bringing with them Frauline (Miss) Selma Close, who acted as maid and cook to them.

Louis Rothchild and Felix Vogel continued their work at the mines for many years. The mine operations were carried on by tunnel, a set of tracks running back into the bowels of the earth, where they were breaking out Turquoise. Several cars pushed by manpower were used to bring the ore out of the mines.

There were eight log cabins located on a hill only a few feet distant from the diggings. In these cabins resided eight Mexican miners and their families. Vogel lived still higher up in a modern adobe, which was built by Mexican labor. Vogel also built a little store, which he ran for the convenience of his employers.

While Vogel was operating this store, a very funny incident happened. He brought Margarine to this mining camp to take the place of butter. W. D. Murray, who operated a store in Silver City, caused his arrest for selling Margarine, which he claimed was poisonous. This was all done through jealousy, because Murray wanted the patronage of theses miners. After paying a large fine, and spending a day in jail, Mr. Vogel proved that Margarine was non-poisonous, and non-injurious, to one's health. After this unpleasant experience, Mr. Vogel seemed to run his store and mine in peace.

About the third year that Felix Vogel worked the Turquoise Mines, his brother, Alfonse Vogel, came to this country. He came to Tyrone to make his fortune, as his brother had, by working in the Turquoise Mines. He not only came there to make money, but he was planning on marrying a girl from New York. He had been engaged to her for some time, but hesitated on marrying her, since she, Miss Lucile Dolge, operated a Felt factory for Pianos, in New York City. Alfonse Vogel put his wedding date off, until such time that he could support his intended bride in the style he knew she had been accustomed. Unfortunately he died with typhoid fever, two years after his arrival to this country, and he was never married to Miss Dolge.

Miss Dolge took the death of her lover very hard and was not married for five years. Then she was on a cruise going to Germany, and she met Felix Vogel, her sweetheart's brother. His wife had died, the year before, giving birth to a child.

By the time Lucile Dolge and Felix Vogel arrived in Germany, a romance had blossomed out for them. They were married and honeymooned in Europe, returning to the United States and to the Turquoise Mines in the mountains above Tyrone, New Mexico, in less than six months.

Louis Rothchild and Felix Vogel had a little office on top of the hill near Vogel's house. There, they sorted their turquoise as it came from the mines. The very best of this fascinating stone was carefully packed in pound boxes and transported to Silver City, by horse and buggy. From Silver City, the turquoise was sent to Tiffany and Company, in New York City. The remainder was sold in Chicago. The best grade of turquoise sold for fifty dollars a pound.

In nineteen and eleven, this turquoise began to turn into copper and Felix Vogel left Tyrone, New Mexico and went to Old Mexico. There, he became interested in a great Salt Mine. At this time, Louis Rothchild returned to England. Only Adolph Armmene remained in this country.

These famous Turquoise Mines have ceased operation, because

the turquoise is turning to copper and they are worthless as a mining venture.

Source of Information: This information was told to me by Selma Close Roach, wife of William Roach. Her address is 102 Pinos Altos Street, Silver City, New Mexico. Mrs. Close was born in Hamburg, Germany. She is 72 years of age.

Hands that Built America: Memorandum on Mining Operations

from the New Mexico Writers' Project

The Conquistadores had hoped to find another Mexico in the Southwest, and the tale of Fray Marcos de Niza and the Seven Cities of Cibola spurred them on. But no gold was found from Cibola to Quivira at first. Not until several years later (1725 according to report), was there any mining in New Mexico.

Tales are told of great discoveries of gold and other rich mines before the Pueblo Revolt of 1680. Tales, too, of how the Indians, after the Spaniards had been driven out, covered all mines and kept the secret of their location inviolate when the Spanish returned after the reconquest. For the most part, these are legends. There did seem to be some reason to believe that there had been a great treasure at Tabira. Jacobo Yrisarri discovered hidden tunnels just as described in charts. But he had not taken out a permit to excavate, and work once discontinued was never resumed. Whether or not the 1,500 burro loads of treasure lay in the direction of his excavations is yet to be ascertained. This gold was supposed to have been mined in the Manzano Mountains nearby. The Indians related that all workmen at the mine were killed, their bodies thrown into the mine, and the mine itself completely obliterated. If there were any mines worked prior to the revolt, the Indians succeeded most effectively in removing any trace of them. It is possible that some of those who retreated with Otermin in 1680 could well have returned to New Mexico with De Vargas twelve years later; more so, if they had knowledge of a rich mine awaiting their return. And of all persons seeking to return, it seems that such a one would be the very one to seize the first opportunity to repossess such sources of wealth as described in legend and folktale.

For several years after the reconquest, few prospectors went to the hills and mountains in search of gold. Though the Pueblos were nominally peaceful, the fierce, nomad tribes east and west of the Rio Grande made forays close to the settlements of the Spaniards.

First ventures in mining employed the crudest of equipment and were successful only where the ore was easily crushed by the arrastras made for that purpose, or on placer ground. Shafts were dug or sunk with much labor and sacrifice. And where shafts were sunk, invariably some pretext was found for using Indians. The mine owners found some way to evade the edicts forbidding the employment of Indians against their will and requiring that they be paid for their labor. Slaves were bought from the Indians themselves. These, they argued, did not come under the edict inasmuch they had been personal property of the sellers and in like manner became the property of the buyer to do with as it pleased him.

The Indians themselves had practiced mining of a sort in their excavations at the turquoise mines near Santa Fe. In contrast to gold, which they did use as an adornment without regard for its intrinsic value, the Indians sought turquoise because it had a religious significance.

The Indians did not excavate deep tunnels in mining turquoise. Their mines resembled burrows, and it cannot be wondered that this is so in view of the stone hammers and sledges they used. Evidences of charcoal and smoke stains on the walls of these ancient diggings have led to the theory that the Indians first lighted fires, heating the rock walls thoroughly, then poured water on the heated walls, causing the lime or rock matrix to crack. In separating the turquoise from the matrix thus dislodged, stone hammers and sledges were brought into play. This method, laborious and slow, testifies to the age of these diggings, for even though the deepest tunnel in them is only two hundred feet deep, it must have taken many years of such painstaking toil to have penetrated even that short distance into the earth's surface.

The Spanish, with iron and iron-tipped tools, succeeded in digging deeper into the bowels of the earth but discovered very few paying mines. They did, however, provide more labor for the Indians, who willingly or unwillingly assisted them. For it was the Indian who carried the ore up the "chicken ladders" in each excavation. These were simply notched poles or logs 12 or 14 feet long leading from one landing or step-platform on the wall of the shaft to another. With a *mecapal* (head band) supporting the rawhide buckets (*tanates*) holding the ore, the Indians climbed these "chicken ladders," which offered precarious footing at best.

It is no wonder that the current belief is that the Indians obliterated every trace of all mines which the Spaniards had worked before 1680. And though the statement is made that there was no mining done in the province until 1725, the persistence of the above belief and the remains found near some old mines, La Mina de la Tierra, for example, in the Cerrillos district, would lead one to believe the opposite. Stories persist concerning the treasures which the padres secreted where they had an opportunity to do so before they took to flight. The instance cited above of the gold buried at Tabira and another current story of the $14,000,000 in gold supposed to have been buried by the religious of Arroyo Hondo and vicinity in the shaft of some rich mine they had been working in the Taos mountains are in point.

Placer mining, the best source of gold in the early days of the province, did not necessitate much or intricate equipment. Any person having a pick and shovel or just a shovel and possessing a *bates* or wooden vessel shaped like the modern miner's gold pan could go to the nearest stream or sand bar and start washing gold.

With the *bates* the dirt, which was thought to contain gold dust or flakes, was scooped up and then submerged in water. A rocking and twirling motion served to free the *bates* of all light earth, which floated away in the onward sweep of the stream. The gold, if any, and pebbles gradually gathered in a segregated group in the bottom of the *bates*. More restrained action, together with the selective discard of obviously worthless stones, left a residue in the bottom of the pan which would contain whatever gold existed. Final flourishes of the pan reduced the contents to a mere handful of black earth or mud among which might gleam several "colors." Quite careful washing now reduced the contents to only the gleaming particles, very few in comparison with the pan full of earth from which they had been segregated. In the old days, these were carefully picked up with the aid of quills, sharpened much as the old quill points for pens were shaped, but with no slit in the nub. Larger pieces, called nuggets, were picked up with eager fingers. A buckskin pouch or small sack served as the universal carrier for dust and nuggets.

Its close texture permitted no escape of even the minutest particle of the treasured dust.

In rich placer country, flumes or cradles were used to expedite the washing of larger quantities of ore-bearing dirt. In both cases the principle was the same. Placer dirt was subjected to the wash of water which carried away the light, worthless refuse, the gold settling down because of its superior weight into little pockets gouged in the floor of the trough or rocker. These little pockets might be either the gouged-out holes mentioned above or cleats nailed at intervals across the width of the trough or cradle. In the recesses thus formed, quicksilver was placed; and this agent, with its affinity for gold, helped to trap the particles of gold carried in the mixture of earth, stones, and water.

After sufficient quantity of earth had been thus washed, operations ceased, and the quicksilver was gathered from the hollows, together with whatever gold had adhered to it.

The collection of quicksilver was placed in a buckskin sack; the sack then was squeezed with much force and thoroughness. The quicksilver oozed through its sides, leaving the gold behind. Gold thus separated was kept until sufficient quantity warranted its melting into a small ingot. This was accomplished by using a clay pot to hold the gold and subjecting it to an intense heat. The molten gold was then poured into a clay mold of the desired size in which it cooled.

Lode mining was rarely done by the early Spanish prior to the American Occupation, and with very little success. The Real de Dolores, near the present town of Cerrillos, is one of the exceptions. Much gold was taken out of this property of the Ortiz family, the original owners. Remains of shafts may be seen on the flats near the old camp site. With no timbering, they must have been dangerous. They penetrate the many years' accumulation piled up in an old riverbed, loose, round boulders and stones with no binding security of soil or sod to hold them. The Spaniards apparently sank these shafts in an effort to reach the old riverbed. Legend has it that in one case they did succeed, and one chamber yielded a rich treasure of nuggets.

Source of Information: From the files of the New Mexico Writers' Project, Santa Fe, New Mexico.

Mining Stories from Las Placitas: Legend of Montezuma Mine

Related by the Whites and Indians in the section around Ojo de la Casa where the mine is located

They say: The old mine was known to the Indians long before the Spaniards set foot on American soil. When Coronado wintered at the pueblo near Bernalillo he visited the mine, which was worked by the Spaniards and the enslaved Indians who were compelled to work long hours with the crudest tools to dig out and smelt the ore for their cruel masters. The gold they dug from the mine was sent on the back of burros to Mexico. When the Indians revolted and drove out the Spaniards in 1680, they filled up all the pits and shafts of the mine. They carried new soil and spread it deeply over all the places they had covered, so that weeds and other vegetation might grow over them and destroy forever all traces of the workings. They carried the rock and ore on the dumps to arroyos and rivers where it could be washed away; there would be no sign left of the mines where they had been compelled to slave.

A Mine for Two Barrels of Water

by W. L. Patterson

The Nannie Baird mine, in the Oro Grande district, in the Jarilla Mountains in southwestern Otero County, a mining property with a production record of over $100,000, was traded by its original owner for two barrels of drinking water. The discoverer and locator of the claim, S. M. Perkins, afterward well known throughout the region as "Ole Perk," was out of water for himself and stock, his water hole near his Three Bears claim had gone dry, so in desperation one day when a wagon came into camp loaded with several barrels of water from the Cox Ranch, twenty miles away, he traded all his right title and interest for two barrels of the precious fluid. The trade stood and from that time onward the Nannie Baird had many changes of ownership.

Perkins was a noted character and well remembered by many of the old timers. It is related of him that he saw while roaming around through the hills the smoke of a camp fire, and blundered accidentally into a camp of none too friendly Mescalero Apaches. Acting on the impulse to slay him for his temerity in invading what they considered their exclusive roaming ground, they first stripped him and thumped him all over. Perkins was somewhat hunchbacked, and that was probably what saved his life. On that account, according to their superstition, the story is told that they decided to spare him, and gave him his freedom. Thereafter he was allowed to roam at will and with prospector's pick and pan worked the Jarilla placers and located a number of lode claims in the district. He became friendly with the Red Men and later acted as interpreter and intermediary for them in their powwows with the Whites.

A little over twenty years after his entry into the Jarilla hills in 1879, in the early 1900s, Oro Grande, having railway connection, experienced a mining boom that lasted for several years until the general slump in mining throughout the West and Southwest. The Nannie Baird became a famous producer and valuable property along with a number of other noted mines of the district.

      Perkins, who came from what was then Indian Territory, now Oklahoma, is reported to have died a natural death in 1916 while engaged in prospecting and trapping in the Burro Mountains near Silver City. He had attained an advanced age, probably around eighty years. Perk Canyon in the mountains a few miles south of Weed where he lived for a time was named for him.

Sources of Information: G. E. Moffett, Mining Engineer and Superintendent at Oro Grande for several years, now of Alamogordo, New Mexico, and Arthur Holmes, Alamogordo, New Mexico.

The Helen Rae Mine

by W. L. Patterson

One of the early discovered mines of the White Mountain region, in Lincoln County, was the Helen Rae, in the Nogal District, discovered in 1865, and first worked by soldiers from Fort Stanton. It is located in Dry Gulch, a few miles below Nogal, once a lively mining camp, but now a quiet mountain village. While most of the well known mines of the district, the American, Rockford, Vera Cruz, Ibex and some others have ceased operations, at least temporarily, for one cause or another, the Helen Rae has continued to be operated, with only occasional close-down periods, until the present day, and only recently is reported to have struck a rich ore body. Other mines of the district have produced good values, but the Helen Rae has been a persistent yielder of pay ore, gold predominating, from the time of its first working.

The first discovery, known as the Cross Cut, was relocated by three "tenderfeet" in 1880, who took out ore of phenomenal richness right near the surface. After changing hands a dozen or more times, the property came into the hands of John Rae, who gave it the name it now bears. "Rae pounded out with mortar and pestle (*New Mexico Mines and Minerals*) from $100 to $300 a day, and in one instance a single pan of the ore yielded $1,100 in gold. A shaft sunk in the Helen Rae 25 feet deep gave $3,300; $10,000 was mortared and panned out, and 125 tons milled $4,804, making a total of $14,804 taken out by Rae in less than a year."

Rae sold the property for $15,000, and it was later acquired by a company which operated it for several years. It consists at present of thirty-seven patented claims, and is leased and operated by John Wall. While known to have a large production, no correct estimate of its total output values are now available.

Sources of Information: *New Mexico Mines and Minerals*, pp. 168-170, F. A. Jones, New Mexican Printing Co., Santa Fe, New Mexico, 1904. A. B. Rose, mining man, Alamogordo, New Mexico (formerly of the Nogal District).

Turquoise Mines Near Oro Grande

by W. L. Patterson

Prospectors in the Jarilla Mountains in the eighties and nineties of the last century found evidences of old turquoise workings in sunken pits and other remains—probably some specimens of the turquoise—left by the early exploiters, most likely the Indians who inhabited the locality, but possibly Spaniards or Mexicans. The workings were shallow and the tools used clumsy, stone hammers and other crude implements found about the old excavations indicating this fact. Charcoal was also found in the working, an indication that fire was used in cracking the stone walls in the process of blasting to uncover the seams containing the turquoise.

About the year 1896, Amos J. DeMueles, a Frenchman and prospector, began the development of a deposit on the site of an old working situated in the hills about five miles a little north of west of the present village of Oro Grande. This development, which came to be known as the famous DeMueles Mine, consisted of several open pits of somewhat irregular shape, the largest forty of fifty feet long and nearly as wide with a maximum depth of about twenty-five feet. Out of these workings it is estimated that DeMueles and his backers produced a good deal more in the comparatively short time they were engaged in the work. One choice specimen, over two inches in diameter, exhibited in a jeweler's shop in El Paso for several years, was valued at $10,000. According to a report another large and valuable stone was sent to France and brought a large price. The untimely death of DeMueles, who was killed by a young Mexican in his cabin on his claim in 1898, robbery being the motive, put a stop to development, and ended the period of production of the mine. After lying idle several years the claim was re-located by J. A. Parker in 1918, but was never afterwards very productive in value of output. At present it is not being worked.

Other turquoise claims were located after the DeMueles, but

only one other went far in the stage of development, the Tiffany Mine, according to report, once owned by the well known jewelry house of that name of New York. This is situated about three miles west of Oro Grande, in a depression between the hills. It is an open pit or cross cut fifty or sixty feet long, ten or twelve feet wide and about eighteen feet in greatest depth, with a side cut of smaller size at nearly right angles from the center. Tiffany ownership cannot be definitely timed, but it was probably anterior to the filing on the property in 1898 by Pat Garrett, noted peace officer connected with the death of the desperado, Billy the Kid, and Mr. W. H. Llewellyn, prominent attorney, one time U. S. Indian agent, member of the state legislature, and well known citizen. They received a patent on the property, known as the Alabama Group, in 1906. They did some mining, as the size of the excavation indicates, but according to reports the production of turquoise was limited under their ownership.

The turquoise is found in seams in hard rock known as Monzonite-quartz porphyry, ranging in size from a mere streak to a seam an inch and a half or two inches or more in thickness. Blasting is necessary in recovering the valuable stones. These deposits are situated in a highly mineralized district containing ores and developed mines of gold, copper, lead and zinc (often in combination) which has produced much value in mineral output.

Sources of Information: *Leading Facts of New Mexico History*, Ralph E. Twitchell, Vol. 3, pp. 447-48, Torch Press, Cedar Rapids, Iowa. *History of New Mexico*, Vol. 2, pp. 971-72, Pacific States Pub. Co., Los Angeles, Chicago, New York. G. E. Moffett, Mining Engineer, Alamogordo, New Mexico, engaged in mining at Oro Grande for a number of years.

Mina de la Tierra

by Robert Pfanner

Examples of Description of Deserted Mine as a Point of Interest to be followed by Field Workers.

Mina de la Tierra, (Mine of the Earth), silver mine, in the Cerrillos Hills. Oldest true lode mine north of Old Mexico, believed to have been worked by the Spaniards with the help of Indian slaves, prior to the Pueblo Revolt of 1680. Tradition attributes the uprising itself to the resentment of the slaves to the cruelty of their masters. The mine has a vertical shaft more than a hundred feet deep together with several side drifts. Ancient coiled pottery and stone hammers, similar to those found at Chalchihuitl, the old turquoise mine in the same vicinity, have been found in the shafts, but until the water has been pumped out no positive statement about the origin of this mine can be made. It is undoubtedly the oldest mine in America, how old no one at present can say. Historians and geologists make no mention of it. Dr. Josiah Gregg, who visited this region prior to 1844, describes one pit of immense depth cut through solid rock. No other mine in the Cerrillos Hills then in existence could have fitted this description. He offers this engaging fact: "Although of undoubted antiquity, these mines have, to all appearances, been worked to some extent within the present century."

Source of Information: *Commerce of the Prairies*, by Josiah Gregg, *The Journal of a Santa Fe Trader*, (New York, 1844).

Mining and Minerals

by Harriet Brent

Mining and Minerals: The history of mining in Santa Fe county antedates the Spanish Conquest. In the memoirs of Alvar Nuñez Cabeza de Baca, who with his companions was the first European to enter New Mexico (1534), there is mention of the turquoise mines in the Cerrillos Hills. Father Zarate-Salmeron also mentions the turquoise mines (see Twitchell's *Leading Facts of New Mexico History Vol. 1*—pages 314-315) but long before the coming of the Spaniards these mines had been worked by the Indians. Together with fragments of pottery of early design, hundreds of primitive mining tools have been found. One ton of stone hammers was recently cleared from an old shaft. It is known that the Indians used rawhide buckets in which to carry ore up the shafts, climbing by notched poles from one landing to another. In the shafts so far explored there is evidence of the only ancient lode-mining in the southwest. One inclined shaft was dug by stone hammers to a distance of at least 60 feet. Ancient and modern methods of mining, side by side, may still be seen, as evidences of the remarkable feats of engineering of the early Indians are still visible. *Chalchihuetl* being an Indian word for turquoise, the mines are known among the neighboring Indians as the Chalchihuetl Mines, and according to the Queres Indians their history goes far back into Queres tradition.

Although at one time these mines were the source of some of the most valuable turquoise in the world, production has now fallen off to a very small output. A subsidiary company of the Tiffany Gem Company of New York once controlled them. After various changes, they are now owned by Los Cerrillos Turquoise Gem Corporation, and are in the charge of G. A. Sewell who lives on the grounds and who for a small admission furnishes guide services for visitors to the workings.

Although no true mining was carried on in the state until 1800, an old document, dated 1713 (Archives at Santa Fe), tells of a covered mine in Old Placer Mt. (the Ortiz group). In 1828 "Old Placer" mine was discovered, and five years later a gold lode was found in quartz veins at the same locality. Dolores and Golden were the centers of activity. (Golden, 38 miles south of Santa Fe on State Highway #10. Dolores, a "ghost town," 15 miles from Golden by third-class road.) The "New Placer" was opened up in 1839 on the north slope of the Ortiz Mts. In veins and in placer gravels gold has been found throughout the entire district south of Cerrillos, and in the basin of the Rio Grande and the Galisteo Rivers. All of the famous old claims, however, were in the Ortiz district.

Lack of water has always been an obstacle to placer operation in the old districts, and in some cases it has been necessary to resort to wells. Gold-bearing gravels are too moist for the "dry" method. In 1900, Thomas Edison erected a plant at Dolores to reclaim rich gravel by a new secret method but the plant was soon closed.

Other metals produced in the county are silver, lead, coppers, and zinc, which with gold came to the value of $2,992,629 between 1904 and 1930. Of this amount, 90% came from the "New Placer" district. Considerable iron ore was once shipped from deposits near Glorieta. Antimony occurs in the Cerrillos district. Gypsum is found along the west border and also east of Stanley. Good clay for brick making is found within the town of Santa Fe, and the Indians have secret deposits of their own for pottery. Manganese has been found four miles northeast of the Capital, but none has been marketed. Mica has been found near Nambe but the mines have never been worked. No oil or gas has been found, although considerable drilling has been done.

The coal mines at Madrid in the Cerrillos Hills employ an average of three to four hundred men who produce from one to two hundred thousand tons annually. The coal crops out on the surface in many places. All shafts are of the incline variety. In production this mine falls below those of Raton and Gallup. Both anthracite and bituminous coal are found at Madrid in cretaceous sandstone and shale. The presence of the hard coal is attributed to an intrusion of molten igneous rock into veins of soft coal, the volatile matter of the latter having been driven off by the heat.

Note: Rewrite by Robert Pflanner and edited by Carlotta Warfield.

> Mogollon N. Mex.
> Sept 23, 1940
>
> Mr. Charles Ethridge Minton
> Santa Fe, N. Mex.
>
> Dear Sir:
>
> Mogollon was founded in 1887. Captain Cooney bought the ~~Modocs~~ Maud S mine and started mining for gold and silver.
>
> It was named for the range of mountains in which it is located. The mountains were named for a Spanish explorer of the early period of Spanish explorations.
>
> Yours very truly,
> Jeff Boone
> Postmaster

(FACSIMILE) Letter from Jeff Boone to Mr. Charles Ethridge Minton, September 26, 1940, NMFWP, WPA #184, NMSRCA

Gold

by D. D. Sharp

The following was obtained from Mr. Thomas Aranda of Palo Amarillo Ranch (Yellow Pole Ranch) in the northeast corner of Bernalillo County, about four miles from the town of Golden. Mr. Tomas Aranda is the son of Mr. Pablo Aranda whose biography is covered by another paper which is attached. Mr. Pablo Aranda discovered the Gold Standard, the Gold Coin, and the Gold King ledges in San Lazarus Gulch east of Golden. He was at the First International Gold Mining Convention at Denver as a delegate from Golden, New Mexico. This first meeting of the convention took place Wednesday, July 7, 1897 and the following is copied from a report now in the possession of Mr. Tomas Aranda.

"In April, 1896 the Gold Coin placer claim was located by Carley, Aranda, and Gallardo in what is known as the San Lazarus Gulch, New Mexico. Immediately a number of men were put to working the placer gravel on the claim and after they had worked for about two months placering, they began to find gold with quartz, which led them to believe that the ledge from which this placer came was near by. They continued dry washing for a distance of seven hundred feet on the eastern slope of the Tuerto Mountains at which place float was found with free gold all through it. A cross-cut was started and the Gold Standard Ledge was discovered at a depth of two feet from the surface. At present (i.e., 1897) three shafts seventy feet deep are being sunk, and a tunnel eighty feet long along the vein has been run, from which one thousand tons of ore have been taken producing $12,000. In working the ore the owners have only run the low grade ore from Huntington Mill at Golden. All the rich ore where free gold could be seen with the naked eye is sacked and saved.

"Besides the Gold Standard Mine Carley and Aranda are interested in two extensions known as the Gold Coin and the McKinly. In the vicinity of the Gold Standard the Agnes Lode, the Golden Chain

Lode, the Comstock, and the Good Enough have similar formation and will be as good mines as the Gold Standard is today (1897) as soon as they are highly developed.

"Messers. Carley and Aranda (Pablo Aranda at the convention in 1897) inform us that in this district there is a scope of ten miles square where placer gold can be found in paying quantities. Several prospect holes have been sunk lower where low grade ore has been found. There is a large scope of country in the immediate vicinity which has never been prospected and only awaits someone with a little push and energy to accomplish as much as Messers. Carley and Aranda.

"We would advise those who are looking for new and rich gold fields to stop at Golden, New Mexico, and call on these gentlemen, who are not only well posted but are thoroughly reliable." (Excerpt from report of First International Gold Mining Convention at Denver, 1897.)

Clipping from Report of First International Gold Mining Convention at Denver, Wednesday, July 7 1897—which is a short biography of Pablo Aranda, now living in Northeast Corner of Bernalillo County, New Mexico, near the town of Golden, New Mexico:

"Pablo Aranda is the son of Mr. Francisco Aranda and was born in Rio de San Francisco (now Golden, the new placer mining district of New Mexico) in the year 1857, on the twenty-first day of January.

"Mr. Aranda has followed placer mining all of his business life and has always been very lucky. He has discovered several placer mines the last one being called Carache on the south slope of the Ortiz Mountains.

"In 1896 he, in company with several other parties, located the Gold Coin placer mines and put men prospecting the same. In a short time quartz with free gold was found, and on September 16, 1896, after six months of prospecting, the Gold Standard mine was discovered, and he had been working it ever since with good results to him and his partners.

"Mr. Aranda has had several mines in the district. He is now part owner in the five-foot Huntington mill where ore is treated at the rate of fifteen tons a day averaging ten dollars a ton.

"Mr. Aranda was married quite young and is the father of thirteen children and was married in the city of Santa Fe to Miss Carmelita Otero.

"Mr. Aranda has always been a friend to everybody in the camp and is well-liked by everyone who knows him. Besides being a good prospector he owns a ranch south of Golden which he farms." End of Clipping.

Mr. Aranda told me while I was at his ranch Friday that he was now working in San Lazarus Gulch and had indications of being quite near a 'lead.' He showed a bottle of gold nuggets which he had recently mined. They were almost pure gold with a small mixture of quartz particles. He says that he ships his gold but that some is sold locally at the general store. He seemed to be quite optimistic that a new and bigger strike was just ahead.

I asked him what was the biggest nugget ever found near his ranch, and he said that the biggest he had ever found was one which sold for $58.00 when gold was at the pre-depression price of about $20.00 an ounce. This nugget he found in a pile of 'tailings' which had been discarded by a dry washer with a screen which threw out all gravel over the size of a silver dollar. He was walking along watching the tailings closely as was his habit when walking over prospect country. He also told me that he heard of a man finding a $60.00 nugget in a similar fashion, but that he did not see the nugget.

At the First International Gold Mining Convention was Honorable H. A. A. Tabor, who is famous in mining history. Mr. Aranda remembered him well. Mr. P. H. Murray of Trinidad represented Red River, New Mexico at the request of the prospectors in that vicinity as he had interests there and was in Denver to attend the convention.

The following from the *Daily Rocky Mountain News* of January 29, 1897, was copied from a clipping in Mr. Aranda's possession. While the prices shown were those at the district named, Mr. Aranda says they furnish a basis for revealing the prevailing prices at northern New Mexico mines which were in some commodities even higher.

"There is every year an increase in the attention given to gold mining by capitalists and investors.

"Did you ever stop to think why gold mining is so very much more favorable than ever before and why hundreds and hundreds of properties can now be made to pay investment profits that could not do so several years ago?

"Below we give a table showing cost of supplies, labor, and treatment in 1870 as compared with the present time (1897). Prices given for 1870 are those in effect at that time in Ward District, Boulder County. Prices stated for 1897 are now in force in the same district. What we say of this district may be stated of the entire State. Some figures may vary in different localities but the average will approximate the same.

	1870	1897
Powder, per pound	$1.00	$0.125
Steel, per pound	.40	.08
Ropes, per pound	.60	.15
Iron, per pound	.25	.03
Giant Caps, per 100	3.00	.55
Fuse, 100 feet	3.00	.55
Lumber, per M	60.00	14.00
Candles, per box	20.00	4.00
Nails, per keg	20.00	3.00
Coal Oil, per gallon	3.00	.20
Wood, per cord	8.00	3.00
Picks, each	3.75	1.00
Shovels, each	2.75	1.00
Miners, per day	4.00	3.00
Teaming, per day	2.00	4.00
Stamp Mill, charges, a ton	5.00	2.00
Smelting charges, per ton	60.00	7.00

"During all these years, gold the product remains the same $20.67 an ounce. With improved machinery and new methods of treatment of ores now in use we can further and truthfully add that a much larger percentage of values are saved than were with the crude systems of reductions of twenty-five years ago."

End of clipping.

La Mina de la Virgen de Oro
or The Mine of the Gold Virgin

by N. Howard Thorp

The Sandia Indian Pueblo grant on which this old mine is situated is bounded on the south by the Elena Gallegos, or Alameda grant. The Pueblo is located three miles south of Bernalillo, on the east side of the Rio Grande, and about a mile from the river, the altitude being five thousand and forty-eight feet.

The Tiwa name for the Pueblo is Nafiat meaning dusty place. The name Sandia was applied in the seventeenth century, and has continued rather than the Saint name of San Francisco, and Nuestra Senora de los Dolores. The Navajos speak of Sandia as Khin Legai, meaning White House.

The population of the Pueblo in 1680 was three thousand people. After the revolt of the Pueblos the Indians abandoned their homes, some of them building a new village on the middle mesa in the Hopi country. In seventeen hundred and forty-eight, the Pueblo was re-established, the Indians having been brought back from the Hopi region. In seventeen hundred and sixty, the number of inhabitants was two hundred and ninety-one, and in nineteen hundred and ten there were but seventy-three persons in the Village.

This grant of land by the Governor of New Mexico, under orders from the Viceroy of New Spain, was made in 1748, of which documentary proof exists. The Sandia grant was confirmed by the Congress of the United States on December 22, 1858, upon the recommendation of the Surveyor General of New Mexico. The following year it was surveyed and contains 24,187 acres; the patent for it was issued by the U. S. Land office in 1864. Sandia was one of the towns of the Province, which in Coronado's time was called Tiguex. Onate, in 1598, visited the Pueblo which he spoke of as Napeya. The Mission of San Francisco was established in 1617 under Zarate Salmeron. At the time of the Pueblo revolt, the Sandia Indians joined in the expulsion of the Spaniards. In the following year, 1681, Otermin burned the Pueblo during his attempt to reconquer the country.

Following this event the inhabitants deserted the Rio Grande Valley, and it was not until 1742 that Friars Delgado and Pino brought the Indians back.

The refounding of the Village on the site of the earlier Pueblo took place in 1748. With the three hundred and fifty Indians who were assembled at this location by Friar Juan Miguel Menchero, the houses were rebuilt and a new Mission was established called Nuestra Senora de los Dolores y San Antonio. The ruins of this Mission church are still to be seen west of the present Pueblo. In 1870 this building was abandoned and a new church constructed to the north of the Pueblo.

I seem to have digressed considerably from La Mina de la Virgen de Oro, though I feel it necessary to give a description of the Sandia Grant, for as I have previously stated it is upon this grant that the lost Mine is located.

Try as you may you cannot get any information from the Sandia Indians regarding this Mine, which they have so jealously guarded for many years past, their only reply to questions being, Si or Quien sabe.

However, an old man now living in Bernalillo, who is some eighty-eight years old, born in Ranchos de Albuquerque, and whose name is Donaciano Gallegos, told me the following story relating to the La Mina de la Virgen de Oro. The location of this old mine is about five miles southeast of the town of Bernalillo, and two miles south of the Canyon del Agua, which runs in a general east and west direction, and is close to the north line of the Sandia grant.

It is a country of washes and arroyos, very broken, and with its scrub timber and brush, hard for one to keep one's bearings in.

Many years ago the stepson of Donaciano Gallegos—named Sarfin—quite a young boy—was herding goats in the country above described, when waking up after a nap, discovered he did not have any goats, or at least none were in sight. Now a goat in one respect is like a sheep, they will always follow their leader. Sarafin took the trail of the goats, and presently came to a large hole; this was on a little neck of land between two small canyons. Peering down into the aperture he saw his goats, nibbling at what little grass and brush the room contained. The top of this room had been covered over with brush and earth, and Sarafin noticed a tunnel to his left hand, also covered with brush, and which extended to the level land outside. As through this tunnel was the only way to get his goats out, he worked long and arduously, removing the brush with which it had been concealed. As he approached the farther side of the room to start his goats out, there on a shelf of rock he saw a large golden image of the Virgin. Sarafin afterwards told his step-father that upon seeing the image, he crossed himself and had started his goats out through the tunnel entrance, when looking up he saw three Indians standing at the edge of the pit into which his goats had fallen. They were furious with him and told Sarafin if they ever caught him around there again they would kill him, and the figure of the Virgin would curse him for ever afterwards. Driving his flock in a run the scared boy appeared at the ranch, and told Donaciano Gallegos of his adventure. Studying over the matter, a few days afterwards Donaciano Gallegos went to the Pueblo of Santo Domingo, and enlisted the help of two Spanish friends to go with him and try to find the mine. A general idea of the location was all they had to go on, as the terrified boy would not return, and several days' rain had obliterated all the goats' tracks. However, fixing up a camping outfit, and packing it on a burro, the three friends started out.

    Arriving at what they believed to be approximately the location, they decided to split up, each one for himself, in this way hoping to cover more ground during their explorations; all agreeing to meet at the camp for the night. Darkness came on and Donaciano and Pedro—one of the friends—arrived at the camp, cooked their supper, and told of their experiences during the day's hunt, but still the other friend failed to arrive. Finally—thinking he would come later—the two friends fell asleep. Morning came, but the third man had failed to show up. They decided to wait for him until noon; then if he did not come they would return to Bernalillo, as the chances were if he had become lost and could not again find the camp he would at least follow one of the many canyons which ran from the mountains west to the Rio Grande, and then strike the main road running into Bernalillo.

Accordingly, after eating their dinner, they packed up their burro, and leaving a forked stick in the ground pointing in the direction they had taken, started on their return to Bernalillo. Where they were traveling entered the main road from the south to Bernalillo; Donaciano and his partner found the familiar boot tracks of the lost man. Arriving at his home, and opening the door, Donaciano—who was first to enter—saw his missing friend stretched out on a colchon before the fireplace, sound asleep.

After supper the lost partner gave an account of his adventures. After leaving Donaciano and Pedro, he said he walked up a canyon—small and narrow—and which ended in a tunnel. Entering the tunnel, he noticed some goat signs, but at the time thought nothing of it. Proceeding along he forced his way through some brush, and entered a chamber in which a large number of Indians were working, busily roofing it over. He was immediately surrounded and several Indians struck him, and then started to blindfold him, but before the bandage was in place he saw at the end of the tunnel a golden image of the Virgin Mary. After his eyes were bandaged, he was led, he thought, through the same tunnel by which he had entered, but away from the mine. As they were all talking in the Indian language, he could not understand what was said, but every little way he was suddenly turned around; this was done so often that he soon lost all sense of direction. They kept hurrying him on, one of the Indians cursing him in Spanish and hitting him with a stick. After they had led him—what seemed for miles—he was told to sit down, and not to remove the bandage from his eyes until told to do so. He waited for what seemed an hour, and then hearing no noises peeked over the top of the blinder; no Indians were in sight. Where he was he had no idea, as all sense of direction had been lost. For many years Donaciano Gallegos has hunted in vain for the lost mine of La Mina de la Virgen de Oro, but cannot locate it.

To be continued.

Dolores:

Dolores, center of the Old Placer distirct, is now a ghost-town of roofless buildings. The name may have been taken from that of the pueblo, Nuestra Senora de los Dolores de Sancia (not the Sandia pueblo near Bernalillo). Gold was discovered here in 1828 by a Mexican while seeking some lost mules. First miners made a wretched living by panning the gravels with water obtained by melting snow with hot rocks. First lodes discovered in 1838 and claim (now in archives at Santa Fe) made to some 60,000 acres, known as Ortiz Grant

After passing through numerous hands this mine grant became the property of the New Mexico Mining Co, present owners. Though the mine equipment has been kept up, little ore is being extracted at the present time, and that chiefly by leasing companies.

The Old Mines Near Las Placitas

Told by N. Howard Thorp

We all know from history that the Spanish Conquerors of what is now New Mexico subjugated various tribes of Pueblo Indians and worked them as slaves in the mines, from which gold and silver were taken and sent to Old Mexico and Spain. The yoke of slavery in time became too heavy for these Indians to bear, and they rebelled, rising up and driving out the Spaniards in 1680. This was the period in which most of the valuable Spanish mines were filled in and covered up, and gold and silver bullion was buried.

There is no doubt that these mines were worked by the Spaniards and subsequently by others from Old Mexico. They are frequently referred to in documents now in the archives of Mexico and old Spanish papers in possession of Don José Gurule of Las Placitas, bearing the date of 1667. The five lost mines in this district are as follows: The Montezuma, La Mina de Ventana, La Mina de la Escalera, La Mina de Nepomuceno, and La Mina de Coloa. The document further states that "Lado oriente de Placitas travagardo La Mina Montezuma Antonio Jimenez" meaning that to the east of Las Placitas, Antonio Jimenez was working the Montezuma Mine, and follows with the statement that Jimenez took twelve mule loads of bullion to old Mexico, and never returned.

The Spaniards and Mexicans smelted the ores taken from the mines in ovens of mud, the remains of several of which now stand and may be seen in the vicinity of these old mines, from which they shipped the gold and silver on pack animals to Old Mexico. By a process of their own—now one of the lost arts—the lead and foreign metals were destroyed, parts of the lead adhering to the slag, as can be seen in the slag piles of these mines near the Spanish arrastras. This process seems to have been simple. The charge of raw ores and fluxes was put into these ovens, and the lead and base metals were destroyed, the gold and silver remaining. These five mines were worked by Indian slaves from the nearby pueblos of Cochiti, San Felipe, Santo Domingo, Sandia, Alameda, and Puaray, the latter being but one league south of the Pueblo of Sandia and the ruins of which can be seen some four hundred feet west of the house of Don Pedro Garcia. These towns are all situated in the Rio Grande valley within a half-day's ride of the mines. Since the time of their slavery, none of the Pueblo Indians of New Mexico, who are very superstitious on the subject, have worked or gone near the mines.

Until these mines had been rediscovered, no amount of coaxing could get the Indians to tell where they were located. While working them, the Indians lived in a village of their own near them, the remains of which now stand in Las Huertas Canyon. It is estimated that about one hundred twenty-five Indian families worked and lived at the spot.

An Indian who some years ago was chief of the Cochitis says his ancestors worked in these mines for six moons in every year, the remainder of the time being spent in raising their crops in the valley of the Rio Grande. According to this arrangement, the different pueblos must have alternated in sending workers to these mines. The Cochiti further stated that the gold and silver ornaments used in the pueblo churches were fashioned from gold extracted from these mines. He also states that the levels of one of the mines caved in, burying many Indians.

Montezuma, also known as Poseyemo, is the male god of these Indians, and it is but natural that the principal mine of this group should have received his name. Indians expect Montezuma to return to earth, to place them again in possession of all their land, and even now at sunrise you can see some of them looking toward the east for Montezuma being borne toward them by an eagle.

The tradition among the natives who now live at Ojo de la Casa is that one of these mines has no bottom. There was no pumping machinery in those days. The water was carried to the surface in earthenware vessels and rawhide buckets on the peon's back, held in place by straps around the forehead of the slaves, who climbed a notched pole, sometimes for a hundred feet or more, carrying the water and ore to the surface. If one should slip and fall with those tremendous loads, all would plunge to the bottom. These people of Ojo de la Casa also repeat the story of the

Cochiti, that the crowns and statues of Saint Joseph, the Virgin Mary, and others in Saint Joseph's Church in Algodones, were made from beaten gold and silver taken from these mines.

It is well known among these people that Felix Samora, while ploughing in a field north of the Montezuma twelve years ago in the vicinity of one of the old ore ovens, brought to the surface a bar of bullion fifteen inches long, one and a half inches wide, and an inch thick, a piece of which bar was tested by a San Francisco assayer and found to be almost pure gold and silver. Five of the old Spanish ovens or ore roasts have been discovered. Recently, a young man named Wilson, while ploughing near the mines, brought up a bar of gold said to be worth nineteen hundred fifty dollars, but unaware of its exact value, sold it for fifteen hundred fifty dollars. These matters prove two things: first, that these mines at one time were extensively worked, and secondly, that very rich ores were formerly taken from them.

The Antonio Jimenez, who took the first mule loads of gold from these mines to Old Mexico and never returned, was from the town of Jimenez in Mexico, the same being named for his family.

Source of Information: *Mining Stories from Las Placitas.*

A Prospector's Experience

by Mrs. W. C. Totty

"A man gets some queer ideas in his head when he's out all alone in the mountains," said John Sanderson, "half of them believe in ghosts, nine out of ten in signs, and all of them in luck. My own experience has changed my views in a good many particulars, and for one thing, it has made me a firm believer in special providences. It didn't come about gradually but through as marvelous escape from an awful death as I believe ever falls to man.

"I had a pet theory then that if you followed the creeks up high enough you would find a tremendous deposit of gold in decomposed quartz. I talked the thing up to Charlie Burk, another prospector and friend of mine, until he agreed to put up half of the outfit and join me in the search. We got a couple of burros, the necessary tools and started early in the spring.

"The country in the Black Range is about as wild and desolate as any on earth, and it was a trip that nothing but faith and enthusiasm would prompt a man to attempt. It was one succession of gorges, gulches, and acclivities, all strewn with granite boulders from the size of a man's hand to a four story block, and often we were obliged to leave the water course that we were following and make detours that took days at the time.

"The creek we followed was almost dry and we stopped frequently looking for placers. We found no very rich ones, but everywhere there was gold. Sometimes there would be lots of it in the bottom of the tin cup after we had taken a drink, and sometimes, here is a curious thing, it would be floating on the surface. I will let someone who is better posted in science than I, tell why gold now and then floats, but I only know that little flakes of it do, and a lot of it is lost in sluice mining that way. As long as we found placers we knew that the main deposit was ahead, so we pushed along, tired enough but confident.

"At last we came to a spot where the sand was barren for several days journey, and then we began to prospect the country around. To make a long story short we struck a ledge one morning with outcroppings that crumbled under my pick and showed quartz all streaked with yellow threads.

"'Charlie,' I yelled, all afire at once, 'we have struck it!' But before we sunk a shaft we found something else that sent our hearts to our mouths. It was an old shaft, back a little and in a claim properly staked out that covered that very ledge. There was a notification according to law on one of the posts, that Peter Summer and Joseph Keautzy had taken possession of the Big Six and done the legal assessment work. I sat right down and collapsed but Charlie went over to the shaft and came back to tell me it didn't cover half the amount necessary under the law to hold the property for the year. We measured it and sure enough, it was down only about one-half the required distance so we took possession of the property, changed its name to 'The Treasury,' and went to work.

"We built ourselves a rough shanty, rigged up a windlass and began to sink. In a few days we were in a formation rich enough to make a man's head swim, and got better as we went down. We were both so excited that we begrudged the time to sleep and eat, and we neither of us meditated for an instant giving the claim up to anybody, assessment work, or no assessment work. What had become of the two men was a mystery. They had left no trace except the notification board and shaft, and it gave me the creeps now and then to think that they might be dead.

"But we were not in a frame of mind to let sentiment interfere with business. I suppose we had been there a couple of weeks when provisions began to run short. We didn't want to both leave the claim at once so it was finally arranged that Charlie would go down the creek about fifty miles to a camp and get supplies. He took the two burros and started off. I calculated that it would take him a week to make a trip, and time hung heavy on my hands. I tried to work a little on the shaft. The formation was very hard and we had rigged up a sort of a cross-bar ladder. I would go down this, fill the bucket, climb to the surface and pull it up.

"About noon of the second day after he left I was startled at what I thought was a man crossing a little gulch a half a mile away. I only had a view of it between two rocks, and whatever it was it passed so quickly that I was not sure. However, I waited for a couple of hours, and then seeing nothing further concluded I was mistaken and I went down into the shaft. I filled the bucket with very heavy ore, climbed up and had it about half raised when a man came walking up the creek bed toward me. Then I knew that I was right before.

"He was an ugly looking customer, big and brawny with a flat Scandinavian face, and carried a Winchester on his arm. I had a little stick that I slipped into the windlass handle near the axle to keep it from turning backward and leaving the bucket just where it was suspended half way up. I started towards the cabin to get my arms. He covered me with his repeating rifle and ordered me to halt.

"'What are you doing on my claim?' he said. 'I reckon you can see,' I replied, pulling as good a face on it as I possibly could.

"'Do you mean you jumped it, you cursed thief?' 'No, I don't, there wasn't enough work on it to hold it, and it was as much mine as anybody's.'

"'You lie!' He looked at me with his wicked greenish eyes for a full minute, then he said: 'Did you ever pray?'

"'Yes,' I faltered. 'Then pray now, I'll give you two minutes to do it.' By that time my mind was clear enough to take in the whole situation, I had no doubt he intended to murder me then and there. With me out of the way there would be no one to testify to the insufficient work, and I would simply be regarded in history when my death was told as claim jumper who had justly been dealt with. I felt my knees tremble and tried another trick.

"'If you kill me,' I said, 'my partner will be back and see that you hang for it.' 'I'll fix your partner the same way, you claim-jumping cur.'

"True enough nothing would be easier than to assassinate Burk on his return, and we had so jealously guarded the secret of our trip that no one would know where to search for us. We would simply disappear, as hundreds of prospectors do, never to be seen by man again, and speedily to be forgotten. I had no hope of mercy from the instant I looked into the man's cruel face. I felt with a sickening qualm and a wild drumming in my ears that my time had come.

"'Oh! For heaven's sake don't murder me,' I cried, 'I will go.' The man made no reply. For a moment my head swam, and then with a sudden return of vision that was excruciating in its clearness, I saw him stoop slightly, rest the gun barrel over the windlass handle, and marked even the slight contraction of the eye-lid that always precedes a shot.

"The next instant there was a crash, an explosion and a cry all mingled into one. I saw the man turning head over heels down the embankment, the Winchester flying through a cloud of smoke up into the air, and all the while I heard a loud, monotonous whirling noise that was like some gigantic clock running down. I did not realize it at the time but this is what happened.

"When he rested his gun on the windlass he dropped his barrel right across the little stick I had thrust in to prevent it tumbling and knocked it out. I suppose the bucket of ore weighed one-hundred fifty pounds, and the great iron handle swinging clear around with such terrific movement, that when it struck him square in the face, which it did, it lifted him off of his feet like a cannon ball. The gun was discharged by the shock but the bullet went nowhere near me. Before I regained my senses I heard the bucket hit the bottom with a smash.

"When I picked up the man he was unconscious, but moaning a little, and the blood trickled on his ears, and his gun was broken. He lay at the cabin for a week or two and after Charlie returned we took him to Silver City. There Dr. Slough put his face in a sort of plaster of paris cast but although the wound healed he was out of his head and eventually died. The night before he passed away he motioned for a little slate he used to write on for he couldn't speak. He was very weak, and it took him a long time but at last he scrawled – 'Who hit me?' Before they could tell him he fainted away.

"I sold my half of the claim a short time after the accident; the mine played out about a year later."

Founding of Silver City

by Mrs. W. C. Totty

James S. Campbell was in the party which first discovered silver at Chloride Flat and laid out the town of Silver City, in 1868. In this party were the two Bullard Brothers, John and James.

Mr. Campbell came here from Colorado with them. There were no R. R.'s in the country, and the country was overrun by Indians, so the overland trip was a hazardous undertaking.

The party originally left Colorado to seek a lost gold mine at the head of the Frisco River, fabulous stories of which they were told by a prospector who had been on the ground but driven away by the Indians.

They found all the signs the prospector gave them to identify the place, but could never locate the mine. While searching for the property they ran out of supplies and became discouraged. They made a forced march down the Blue to the present site of Clifton, where they obtained food. They then struck east into Grant Country, came through the Burro Mountains, and reached the present site of Silver City in the spring of 1868.

Mr. Campbell said the flood marks of driftwood on the trees was in some instances twenty-feet high. He tried to discourage the men from laying out a town site in the canyon, to only have it swept away by future floods.

When offered his choice of a block of the town site, Mr. Campbell said he didn't care for a block of mud holes, as the site then appeared, and refused the offer.

Bullard Street, the main business street of Silver City, and Bullard's Peak, the lofty imposing mountain west of town are named after the intrepid prospectors and pioneers, one of whom gave the town its existence, and who later gave his life while protecting his property from hostile Indians just west of town.

His body now lies in the City Cemetery, the first white man to be buried in Silver City.

Source of Information: James S. Campbell, Published in *Enterprise*, Oct. 21, 1910.

Gold Gulch Findings

by Frances E. Totty

Many years ago two soldiers from Fort Bayard, when they were discharged, went prospecting in the present Gold Gulch country. The men struck a fairly rich claim and after they had gotten together what was considered a small fortune they decided to return to their childhood home in Iowa and there bought farms and later married and raised families. These men always planned to come back to this place where they found their riches, but it seemed that something always kept them at home. Finally when two of their sons were grown, they decided to return to the claims of their fathers and on their arrival in Silver City they journeyed into the Gold Gulch country, and located the claims that the two soldiers had staked off and everything was left as it had been when the two men left the country; the map was accurate and the stakes were as the men left them.

The two youngsters now knew that they were soon to have a small fortune as their fathers before found out of the same place that their fathers got their start, but imagine their surprise when they discovered that ninety percent of the ore had been removed by the older men when they had stopped and decided to return home to Iowa. The two youngsters remained and tried for some time to find riches in the Gulch, and as the men had worked the claims that their parents had staked out they returned home disappointed, but rejoiced that their fathers had been able to get most of the gold from the Gulch.

Mining Life

by Frances E. Totty

Mining Life, founded by O. L. Scott, the first issue of which appeared in Silver City on May 17, 1873, was the first newspaper ever published in Silver City.

Mr. Scott, in making his bow to the public as he entered the journalistic field here, called the attention to several handicaps that burdened the growth of Silver City. Among them was the following:

"Our isolated position, and the fear of the dreaded Apaches, has heretofore been a bar to immigration, but owing to the vigorous measures lately pursued in Arizona we can now offer to the immigrants a peaceful home, be he inclined to develop our exhaustive mines of silver, copper and gold. To follow the plow in our fertile valleys, or watch his flocks fatten, summer and winter, on our nutritious grasses, satisfied that this country can offer a wider range and freer scope, in which to follow his bent, than any country in the west."

The following articles were taken from the *Mining Life* of 1873:

"Indian rumors—A great many reports have been flying about town the last few days in regard to Indians. Last week quite a commotion was raised by the rumors that a boy had been killed near town while herding stock. A party of gentlemen armed themselves and proceeded to the spot, but found the statement false. Again on last Monday, it was reported that a large band of Indians were seen the previous night in the vicinity of Bramen's mill, and that they had killed two oxen nearby. We have been unable to trace this report to any authentic source. Our citizens are ever ready to assist those who are in distress from the Indians or otherwise, but their patience is becoming exhausted by these numerous embellished statements of enthusiastic news mongers who seem to be gifted with a immense fund of the 'wonderful'." *Mining Life*, 1873

"Indian raid—During the whole of last week rumors were afloat that Indians were seen in the vicinity of town, and killed three cows from Pinos Altos. On Monday, last week, Charles Brakebill, with a four-mule team and wagon, loaded with lumber, were coming in from Bremen's Mill, and when in three miles of town, was attacked by seven Indians, six on foot and one on horseback. Charley having his hands full in keeping his four frightened mules on the road, he could not pay much attention to the Indians but gave them the best he had with his revolver, and only abandoned his wagon when the Indian on horseback in the rear called to his comrade to head him off on the left; another commenced firing from in front with arrows. Charley came into town on foot, minus hat, but with empty pistol in one hand and a blacksnake whip in the other. Word was sent to Capt. Fechet, at Mangua springs, and we hope that with so early a start as the Capt. will be able to make on the trail, that we can have a scalp dance on the Fourth." *Mining Life*, 1873.

Source of Information: *City Enterprise*, Aug. 31, 1928, by Agnes Snyder.

Negro Findings

by Frances E. Totty

In the early '70s the Indians were causing quite a bit of trouble in the Animas country. The Indians had caused so much disturbance that the soldiers from Fort Bayard were finally called out to try to put a stop to some of their degradations and mischief. The soldiers at the Fort at this time were negroes, and they didn't care too much to encounter the Indians, but when they did as a general rule they gave them a good fight. They trailed the Indians into the Animas country about where Victorio Park is located today. The negroes made camp close to a waterfall on the side of a hill. One of the soldiers picked up a nugget while making camp and another was found by the waterfall.

Just at dusk the Indians decided to raid the camp for the horses that were along and a fight followed with all the soldiers killed, but one that hid under the waterfall. After the fight was over and the Indians had gone on their way, the soldier that had hidden under the falls began to look around and found a much larger nugget where he had been standing. After picking up several nuggets he began to try to make his way back to camp and after some time of wandering and living on herbs and wild animals he did get back to a ranch and was directed to the Fort. He tried to tell where he had found the gold and led several parties to the district, but as far as is known the place was never found. It seems to lie between McKnight Canyon and East Canyon to the best any can decide, or was the negro farther up in the mountains and in his wandering lost his bearings? This mining place may some day be found, but again it may be another Adam's lost mine.

Source of Information: Leslie K. Goforth.

Field Notes   Jordan, Mrs. Mildred

NOTES ON PINOS ALTOS RANGE

Location: Eight miles northeast of Silver City, New Mexico. Situated on Continental Divide. Bear Creek water runs to Pacific, Whiskey Creek water runs to Atlantic.

Name: Range and town named Pinos Altos-- meaning "high pines". Former name of camp was Birchville.

Mining: Placer gold discovered in 1859 or 1860. Mining short duration, because of Indians. Gold, silver, copper, lead, and zinc.

Remarks: Richard F. Clark, old soldier, relates many early happenings with the Indians. In February, 1871, He was wounded by the Indians while carrying mail from Fort Bayard to Pinos Altos. Jim Taylor killed by Indians. John Bullard killed by Indian. The same time the Indian shot him, he shot the Indian, both dying at same time. Shooting done under tree in Silver City. Captain Kelly asked Mr. Richard F. Clark to play his instrument and Billy Cummings a drum for funeral.. Billy became too drunk to play and they had no music.

Pinos Altos: The Great Gold Producing Camp of the County

by Frances E. Totty

"I love the soft mellifluent tongue,

That from the lips sweetly flow

Like the strains with harp and timbrel sung;

The Language of New Mexico."

The first tangible discovery of the precious metal (gold) was made on the 18th of May 1860, by Messers. Snively, Birch and Hicks, three adventurous and daring prospectors, who outfitted at Mesilla, and on their journey westward replenished their supplies at Fort McLean (Apache Tahoe) and afterward at Santa Rita. Pursuing a westerly course from the latter and last outpost of civilization they forged ahead and on their second day out, Birch, the leader of the party, discovered free gold in Bear Creek while in the act of drinking water from the stream. The site of the discovery is in the near vicinity of the Mountain Key mill, just above the junction of Little Cherry Creek with Bear Creek. On making known his discovery, the ground was prospected and the value and extent of the placers ascertained as far as the means at hand would permit.

Returning to Santa Rita, ten miles distant for supplies, the news was confided while in route to the Marston Brothers and Langston, who were then in the employ of Leonardo Siquieros, a lessee of the Santa Rita copper properties. Returning to the discovery it was christened Birchville, in honor of the discoverer. The month of December following saw over seven hundred men in the new field, all more or less actively engaged in the washing of gold in the gulches and arroyos tributary to the Bear Creek. The first year was devoted exclusively to placer mining, and not until 1861 was any attention given to quartz mining. This year the Atlantic and Pacific mines were located and the surface ores ground and amalgamated in arrastras.

Tom Marston, alone (was) running ten tons of ore from the Atlantic. Aside from location nothing in line of development was done on the Pacific. Then summer noted the discovery of the Locke lode, now Mountain Key, which was mined principally for the chipas or nuggets obtained from the surface quartz.

At eight o'clock on the morning of the 22nd of September a large force of Apaches numbering about four hundred, under the immediate command of Cochise, attacked and made a bold but unsuccessful attempt to drive the settlers out of the country. Captain Marston of the Arizona Scouts C. S. A. consisting of nine men, quickly rallied his troops, and the miners rapidly placing themselves under his command, the fighting became general. The country was heavily timbered with little or no underbrush to impede or retard the movement of friend or foe. The Apache right rested along and near the crest of the ridge of the present town site and extended in a northern direction a distance of one-half mile to a point near Skillicorns' mill. The main and hottest part of the engagement occurred in this vicinity, and from the opening of the fight until the close at one o'clock p.m. every foot of the ground was stubbornly contested. At twelve o'clock Marston fell mortally wounded, and died a few days after. At one o'clock the Apaches retired with a loss, it was afterward ascertained, of fifteen of their warriors. The loss of the whites was three killed, including Marston, and seven wounded. During the fight a dog belonging to Carlos Norero grappled with and succeeded in killing an Indian and ever afterward was considered the hero of the day. The following day the whites almost deserted the country, some going north to join the union forces and others casting their fortune with the Confederacy. On the White water the fleeing miners were again attacked by the Apaches, presumably the same band and corralled for two days, most of the time being without food or water. Couriers were sent via Santa Rita, thence to the Mimbres and Lieutenant Swilling of the Arizona Scouts came to the relief, and the party proceeded to the Mesilla valley. Santa Rita, San Jose, and Hanover mines were abandoned shortly afterward.

From the opening of the Civil War up to present time, 1861, Birchville has been known as Pinos Altos, so named by the few remaining Mexicans and signifying "tall pines."

Source of Information: *Silver City Enterprise*, January 3, 1890.

Silver City Mines

by Frances E. Totty

On the 10th day of May 1870, a party of men consisting of James B. Bullard, Joseph Yankie, Elijah Week, H. M. Fusion, James Campbell, a Mr. Webster, Albert Fields, and Joseph Kirk were returning to the mining camp at Pinos Altos from a prospecting trip through the rich copper region situated on the Francisco river in Arizona. They camped on the Cienega de Vicinte, where Silver City now stands, near the site of the Wisconsin Mining company's mill. Three of the party, Bullard, Yankie, and Weeks, thinking they had found some rock that contained mineral took small pieces to Pinos Altos with them and had an assay made by Mr. Johnson which yielded $60 in silver to the ton of ore. Thinking they had struck it rich, as the ore they had assayed was picked up on the surface, these three returned in about a week and located N. E. and Black Hawk ledges.

Doing the stipulated amount of work to secure a claim for a year they returned to Pinos Altos. Taking pieces of the ore for assay they found quite an increased yield over that from the surface and the fever from the excitement over a new discovery spread quite rapidly. We remember that in June, 1870 a piece of ore was shown to the writer at Fort Sheldon, which was said to assay fifty cents to the pound and only ten feet of the surface.

About the middle of June the same parties returned to the Cienega and erected a cabin near the spot now occupied by the Miller and Skillecorn's blacksmith shop, and commenced work on the Legal Tender ledge, sinking it to the depth of about ten or twelve feet.

With the incoming hordes came Mr. Cerasco, and with him some Mexicans from silver mines of Mexico; after looking over the country these parties found and located (the mines) in the Providence ledge, situated in Chloride district. They erected small furnaces such as they used in Mexico for smeltering of silver ore, and succeeded in making a paying business of mining and smeltering, sending to market the first bullion produced at the camp.

During the early days M. W. Bremen had his sawmill located about ten miles northeast of Fort Bayard. On the 10th day July, '70, the Apaches made a raid upon his stock and captured the major portion of them; but in no way daunted by this, as it was but a repetition of an old game, he procured other stock and moved his mill to a point within five miles of the infant town and furnished lumber which enabled the miners to construct habitable houses. The camp now struggled along with varying success: today a small shipment of bullion that would cause a prospector's heart to swell with hope; tonight the lurking Apaches would steal his last burro or ox and leave him alone with no means to move around, save afoot, with his blankets, tools, and provisions on his back. The news of the richest of the ore now spread abroad, and miners came in from the Moreno mines, Colorado, California, Nevada, Montana and even from White Pines came the lusty but hungry miners singing:

"Two hundred feet on the Eberhardt, White Pine Johnny," etc.

On the 15th day of June 1871, Messers. Whitehill, Tidwell and Simpson commenced producing ore by smelting, but owing to the want of good material for furnaces they could not make it pay. They then tried the Patio process and obtained $28.00 per ton. They next took a whiskey keg and tried amalgamation by revolving it on axis. After the first trial they obtained from 75 pounds of ore $64. Quicksilver being scarce they had great difficulty in obtaining a supply, often paying as high as $2.50 per pound. During part of this time they obtained with two arrastre and two barrels $110 per day. In the month of February they sold the "Mud Turtle" to Messers. Arnold and Webb. But during the time they owned it they treated about one hundred and fifty tons of ore and reclaimed about $14,000.

Messers. Arnold and Webb commenced treating ore from the "Mud Turtle" in April 1872 and ran about two and a half months, when the broiler gave out and they were compelled to suspend operations. Then they treated ore by amalgamation, and received as results about $16,000.

Seeing the success achieved by the rude processes around him,

Mr. Bremen purchased a worn out gold mill of eleven stamps, at Pinos Altos, and moved to Silver City in the month of September 1871, using as amalgamators wooden barrels, the same as those used in the "Mud Turtle" up to December 1872, when he purchased and put in a new Wheeler pan. By this time the stamps and batteries, nearly worn out when he got them, were so far gone that with the utmost ingenuity he could only make five stamps available and with these kept on until March 1873, when he concluded to stop and order new pans anxiously awaiting the arrival of these new necessary parts to start up again. During the time the mill was running Mr. Bremen informed us that he treated 400 tons of ore, principally from the Seneca mine, and received as a result about $57,000; and that during the time he lost among the tailings over 60 barrels of quicksilver, showing how wasteful must have been the treatment. John and James Bullard, John Swishelm, and Joseph Yankie put up arrastres in November 1871, using barrels for amalgamation until October 1872, when in conjunction with Col. Rynerson, they purchased an old gold mill for use in Pinos Altos, consisting of five worn out stamps. With these they worked a kind of hermaphrodite pan, made of odds and ends for the purpose of amalgamating until the latter part of January 1873, when on account of the mill being worn out they had to stop work.

During the time they were at work Mr. Bullard informed us that they treated about 220 tons of ore from which they took about $25,000. The ore treated was principally from New Issue and Dexter mines.

The Wisconsin Mining company was organized by Mr. Coleman, who first came to the camp in December 1871. He looked over the country and departed, professing himself satisfied. He organized the company as it now exists and brought the machinery consisting of a Drake crusher and two pans of antiquated pattern. The company, through Mr. Coleman, purchased the Two Ikes mine, situated in the Chloride district. From this mine they have kept the mill constantly employed, crushing and treating on an average rate of $60 per ton, which for fourteen months makes an aggregate of approximately $80,000.

The machinery of the Cibola Reduction and Smeltering Works company, consisting of six stamps, rotary battery and one two-stamp prospecting battery, two settlers, two concentrators, and four large amalgamating barrels, arrived in July 1872, but was not in good operation until January 4, 1873, since which time, although it has not kept steady at work, has treated about 200 tons of ore, principally from the Two Ikes and Providence mines, producing about $12,000.

The smelting works of Mr. Carasco are still in operation and we are informed by Mr. J. R. Johnson that during the entire time since starting, they have treated about 300 tons of ore and having produced about $200,000, thus placing them in the front ranks in production as well as the pioneer bullion producers of the camp.

We can thus sum up the gross productions of the camp at over $330,000, thus produced under the most discouraging circumstances and under the most primitive appliances. This may not seem much to those who have railroad facilities and mines developed, but taking into consideration that we occupy the one point, in the United States, furthest from railroads and all facilities for accomplishing results, we claim that it speaks volumes for the camp.

There are two mills of ten stamps each, with complete appliances for the reduction of ores, one now complete—that of the Tennessee Reduction company and the Pope mill, the stamps of which are in place; but the amalgamating machinery has not yet arrived, but is en route and will arrive as soon as possible.

The song of the "Turtle" is hushed, but as we go to press we hear the thud of the ponderous stamps of the Tennessee Reduction company's mill, and each is anxious to hear the results, recognizing the fact that it is the most complete mill yet brought to camp, and miners are sanguine as to its performance. And each one who owns a mine says they have from five to a hundred tons of ore ready for treatment. Heretofore the difficulties have been that miners could not get their ore treated, consequently they could not develop their mines. So far the camp has been self-supporting and the prospects are brighter than ever and each one meets you with a smile full of hope, feeling confident that if work is done, success is sure to follow.

Source of Information: *Mining Life*, 1873, written by O. L. Scott.

Snakes in a Mine Shaft

by Mrs. W. C. Totty

"Talking of snakes," remarked Col. Richard Allen, former foreman of the Pauline, "we have a little experience with one of them at the mine which some of the boys will never forget." A shot was fired at the bottom of the shaft, nearly seventy feet from the surface, and after the smoke had cleared away a couple of miners were let down in the shaft.

When they had reached the bottom the men at the windlass were startled by terrific screams from below, among which they managed to distinguish, "For God's sake, haul us up!" The men at the windlass pulled for life, thinking that the shaft was caving or that it was full of gas, or some calamity of that sort.

When the miners reached the surface they were nearly dead with fright. The cold perspiration stood out on their foreheads in great beads, and they didn't look as if they had a drop of blood in their veins.

After a while one of them said "Snakes" and finally they recovered sufficiently to say the whole bottom of the shaft was one wiggling, hissing mass of rattlesnake.

A miner was then sent down with a stick of a giant and fuse which he lighted and dropped down among them and then was hoisted.

After the giant exploded 27 rattlesnakes were taken from the shaft, some of them having ten to twelve rattlers on their tails. It seems that the shot fired had opened a little cave which was full of snakes and they had all fallen into the shaft. The boys were rather nervous for sometime afterwards.

It was rather remarkable that the two boys who went into the shaft were not bitten, as the snakes were striking at them from all sides.

Source of Information: Col. Richard Allen.

12/ 4/ 36
Jordan, Mrs. Mildred
Mi.2, cl. 200 words

DEC 7 1936

DESCRIPTION OF AN OLD GHOST TOWN
GEORGETOWN

The district north of Hanover gulch is where the developments of copper and iron properties made a big boom in silver in the seventies. The discovery dates back to 1866, the center of the boom was at Georgetown, at bthis time one of the greatest silver camps in the West. The Maiad Queen, the Quien Sabe and the Commercial and Silver Bell, were prominent mines, but are now idle.

A visitor relates: "In entering the town of Georgetown late in the afternoon of April, 23, 1903, that he and his companion were quite depressed by the awful stillness that prevailed the premises. They found nothing doing. The streets just ghostly shadows and grown up in weeds. Long row of buildings, casting shadows by the lingering sun, causing a feeling of awe and horror. Once the town had been alive with a hustling, moving throng of sturdy prospectors and miners who had " struck it rich". The two men felt the utter desolation, realizing a passing of a western mining camp. But at the end of these series of desolated regions they beheld, towering above the wreckage and piles of waste, a beautiful monument of solid silver, shining in the sun, representing a production of $3,500,000 in credit to the camp".

Source of information: New Mexico History, Vol. II.

This town is on the Black Range Highway

The Founding of Silver City

Taken from the Mogollon Mines: Published in 1914

by Frances E. Totty

During the Ralston silver and diamond excitement in 1871, the travel from Santa Rita, Pinos Altos and Central, along the eastern base of the Burro Mountains to the southern portion of the country, assumed the character, and frequently the proportion, of a stampede. The northern, southern and eastern slopes were not prospected closely, copper at that period held no place in the estimation of the prospector, either in thought or vocabulary. The discovery of silver was his greatest ambition, and only when the urgent demands of business made it obligatory, was his attention directed to the great and unparalleled mineral wealth of the Burro Mountains southwest of Silver City. The business requirements and demands of the newly discovered mining localities adjacent, within the limit of the new mining metropolis on Legal Tender Hill (one mile southwest of Silver City), Chloride Flats (adjoining Silver City to the west where silver was mined by modern means in this district), was discovered by Jim Bullard and his party in 1871. The discovery of rich ore at Lone Mountain, and later at Georgetown (southwest of Silver City, sixteen miles, discovered by Frank Bisbee), required and demanded great personal sacrifice on the part of every individual engaged in the extremely hazardous vocations of prospecting, mining, lumbering, and ranching, to venture one mile distance from town in any direction alone, as the warring Apaches were everywhere. Notwithstanding the daily momentary perils the pioneers fought and forged their way to the Burro Mountains. The first actual settlement was made by Col Joseph Bennett; deceased in 1873, Bennett was one of the founders of Silver City, who established a saw mill on Cherry Creek near the present site of the King Wade's Ranch. The first house and home in the Burro Mountains was built for Richard S. Knight in 1872, near the head of Knight's Canyon. The building was erected by Honorable Robert Black, an early settler of Silver City from Fort Bayard, deceased in 1874. Knight's ranch and canyon soon became a household word throughout the southwest, and on more than one occasion was a haven of safety to the prospector and traveler, who were forced to seek the protection of the ranch as against the attack of the murderous Apaches. Knight's Ranch completed, furnished a cordon of settlements from Las Cruces via Santa Rita, Hanover, Fort Bayard, Pinos Altos, Silver City, Cherry Creek, Knight's Canyon, and Ralston. This led to the establishment of a mail route to California; this was followed by way of Arizona; this was followed three years later by a telegraph line, also built by the government, constructed by Lieutenant Vedders and Reeds and by Sergeant Max Frost to Shakespeare. The Apaches regarded the setting of poles strung with wires with great disfavor, and they were an unwarrantable intrusion on their domain, and they resented it by attacks from ambush on the settlers, whom they tortured, mutilated and murdered, and ravished and murdered the female members of his family. There is no instance, historical or legendary, instancing that the Apache ever tried conclusions with his enemy in the open, but invariably accomplished his deeds of violence and bloodshed through stealth and ambuscades (as to this statement there is no proof as I could find).

In 1877 Colonel Bennett disposed of his saw mill to Messrs. Black and Cosgrove, two business men of Silver City at the time. During their ownership of the equipment of logging teams and wagons, a woman, the wife of one of the employees, was the proud possessor of a stove, the only one within a radius of twenty miles of her tent. The Indians were committing depredations daily, and she had been warned repeatedly to move into camp, as the location of her tent was not only a menace to herself, but to all of the camp. As often as warned, she refused to move, and on being questioned closely, she replied that if she "left for only one-half a day the Indians would come and break her stove all to pieces so they would." The sawmill men quietly picked her up and in spite of her protest, physical resistance and intemperate language, carried her to a place of safety. A day later within three miles an ox train of wagons laden with provisions consigned to John B. Morrells, of Silver City, was attacked, one driver killed, several oxen butchered and mutilated, and the contents strewn over the ground, including the flour. For the better protection of the settlers a block house was built in Sept. 1877 at Silver City, and for one whole month the settlers enjoyed a short respite and much needed rest from incessant raids and warfare. But all too soon the Apaches were to cut the adobe walls of the corrals with lariats and water, and climb over the walls of the block house.

During the month of October, waiting for an opportunity for murder without incurring any personal risk, the enemy bided his time and waited for his opportunity. R. S. Knight and A. B. Connors had contracted with the government, and were delivering hay to Fort Bayard from Whitewater. On their return to the Knight's ranch, a little after sundown, they discovered that they had lost a nut from one of the axles of their wagon. After an unsuccessful search of an hour or more for the nut, the teams were unhitched, and rode into camp. At daylight the following morning, A. B. Connors, John Connors, his son, and a Mexican left for the wagon to bring it into camp. As the three men approached the wagon, which was undisturbed, and as they had left it, they were fired upon. The elder Connors was instantly killed; John Connors mortally wounded, and the Mexican shot through the arm. The Mexican escaped and gave the alarm. John Connors, after he fell, crawled to the remains of his father and tried to bring them away to prevent mutilation, but a second volley compelled him to desist. The elder Connors was found as he fell, but the younger man was not found until the second day, and was partially hidden in the tall grass.

Source of Information: Taken from the *Mogollon Mines*, published in 1914.

Turquoise

by Mrs. W. C. Totty

One of the largest turquoise mines in the world is situated in the Burro Mountains of Grant County. From Persia comes turquoise of fine color and exquisite polish, but no single mine in the country has produced gems of a finer grade or in such abundance as have been taken from the mines in the Burro Mountains.

Turquoise is found in Colorado, Nevada, Alabama, California, and Arizona, but the mines of New Mexico since their discovery in 1890 have furnished the trade in the United States with more than two-thirds of its wares because it is here that the purest gems have been found within the last 50 years.

The turquoise has been used as an ornamental gem from the beginning of historic time, and to many the finding of stone implements near old workings of the turquoise property is conclusive proof of its use in prehistoric days. Small openings have been found with black walls as if smoked, as well as other indications that this mine's district was worked before.

The Mohammedans used polished blue stones in their buildings of worship; the Peruvians ornamented their temple of the sun with the gems and in one of the principal halls of justice in the old royal palace of Tezcuco, in Mexico, was found a throne of pure gold inlaid with turquoise.

The Indian name for turquoise, "Chalchihuitt," was given to a mountain near Santa Fe, where the stone is still found. The archives of Santa Fe record the Chalchihuitt grant bearing the date of 1763, and a broad traveled road leading from Santa Fe to Silver City, to the City of Mexico, would appear to indicate that many of the gems used for ornament in the royal palaces of the Aztecs were brought from the Cerrillos and Burro Mountains mines.

The mining of turquoise in these places was undoubtedly carried on long before the Spaniards came to America. Immense depressions indicating excavations and old dump relics of the age of stone implements are found, constituting the proof.

Hammer heads of flints, bearing grooves for the binging on the handles with which to wield these clumsy implements are found. Coiled pottery of the oldest known type is also found near the workings.

The methods of the prehistoric workers were crude and slow; with the hammers weighing from 1 to 20 pounds, the stone was patiently chipped away. There are some slight indications which would lead to the belief that fires were made upon the rocks and water thrown upon the hot rocks in order to crack it, and thus release the turquoise, but this theory lacks proof since the fire markings cannot prove their own antiquity.

In the Burro Mountains many such old implements have been found, proving that the stones were polished where they were produced by rubbing them upon sandstone of graduated sorts from the coarse grain to the finest. Small stones of the shape that is readily fitted into the hand are found with highly polished grooves of perfect curved shape, showing the final perfection of the gems ground by hand.

In the order of their modern discovery the locality of the New Mexico turquoise mines is as follows: Los Cerrillos, Santa Fe County, Burro Mts., Tyrone, Grant County, Hachita, Grant County, and the Jarilla Mountains in Otero County.

The mines which have produced the finest turquoise are the ones in Grant County: The Porterfield Turquoise Mines Co. and the Azure Mining Co.

The modern discovery of turquoise in the Burro Mountains was made by an old prospector by the name of Nick—who was grubstaked by W. C. Porterfield, a druggist.

"Where are you going, Nick?" asked Mr. Porterfield.

"Oh, I saw some stuff the last time I was in the Burro Mountains, and I'm going back to investigate," replied Nick.

A few weeks later Nick came in with a small rag full of stones and placed it on the counter.

"What you got there, Nick?" remarked Mr. Porterfield, "copper?"

"No, a stone called turquoise, I think," replied Nick.

One of the Porterfield brothers sent the specimen to New York to be assayed. A representative of Rothchilds of Maiden Lane came out and was sold two claims for $2,500 each. They started producing at once under the name of Azure Mining Co. When the Porterfield Mining Co. got their mines to producing they had no market only local until they received a contract from Albert Larson and Co., competing Co. on Maiden Lane in New York to Rothchilds. The Porterfield Mining Co. holds medals from the Saint Louis Exposition, Chicago Exposition and others for their gem display. When the demand for turquoise disappeared, the Porterfield Co. sold to the Phelps Dodge people.

It is said that the appearance of turquoise on the Indian claims was the first indication to the modern masters of the land that turquoise could be found in the vicinity of Cooke's Peak. Cooke's Peak was Geronimo's stronghold and the Apaches raged over the Santa Ritas, Mogollons and the Burro Mountains as well as the Black Range.

Their pottery, ornaments, and weapons are found everywhere. Almost every claim picked up or bought from the Indians of historical value contains at least two messages. One comes from the form of numerous shells which speak of the day when the southwest country was under water and the other in bright polished blue beads which have no counterpart save in the gem of turquoise. Turquoise mines are not worked with precision or rule as the precious stone is variable in its occurrence. Sometimes one blast of dynamite will open a pocket of valuable stone, and again 100 feet or more will be worked without discovering one bit of stone of commercial value.

Turquoise of the finest color and hardness is more rare than diamond of the same grade. In the whole city of New York probably not 20 pieces may be hoarded by the dealers, but on the whole the deep blue azure turquoise is not often seen because it cannot be found. It has been something like 45 years since the famous Elizabeth pocket was found. The turquoise found in this pocket occurred near the surface and was of a blue unequaled by any ever found in Persia and finer than any in America. It was sold for $20 per carat wholesale. It is estimated it brought $500,000,000. It would have broken the market to produce this quantity of flawless gems at any one time and without doubt the stone was carefully hoarded by the dealers in New York and gradually put on the market. Little of this grade of turquoise can be bought now.

The turquoise is found in seams, pockets, lenses, and nuggets. The finest turquoise is usually found in nugget form, a lump bright blue imbedded in the mother stone.

The interior of turquoise is not without beauty.

Much turquoise fades very rapidly a short time after being exposed to the air, and is comparatively valueless.

Water is so inherently a part of turquoise that a prolonged immersion will recharge the gem with a desirable dark blue color, but the effect is not lasting. Other qualities of the stone are affected by light and most of them must be kept some years, preferably, in order to test their retaining capacity.

The turquoise is chipped out of the rocks, or picked up in nuggets as it occurs from the blast.

The production of the turquoise matrix is given as much study as the flawless gem. There are three kinds of turquoise matrix: that which is veined with iron, that which is half rock and half turquoise, clearly defined, and that which is mottled a darker blue spot on a lighter blue background, or vice versa. This latter kind is often beautiful, and sells for $1.25 a carat wholesale; the price is a remarkable contrast to the Elizabeth gem that sold for $20.00 per carat. The mottle turquoise matrix is generally found to contain about 50 per cent pure turquoise. For the last 50 years this grade has been used for ornamental jewelry.

Turquoise has a use quite aside from its setting as a gem. The decorative value is unquestioned, and when the stone is very hard it is capable of workmanship. Turquoise inlaid with other precious stone was greatly valued by the ancients, and few beautiful instances of their art remain.

Mr. Rothchilds of London, many years ago came into possession of a single exquisite turquoise vase. It was about six inches high and reputed to have come from Persia. The color of the vase was the deepest blue and it showed a brilliant polish.

Mr. Rothchilds earnestly desired to have a companion vase to match the Persian production, and for many years searched in vain for a sufficiently large piece of the gem to produce a similar deep blue vase. It is said that he had men searching in Persia many years, and finally turned to America as a last resort. After searching through the various turquoise producing properties, one of Rothchilds' men came to Silver City where he heard of the old Indian workings. One day a Mexican came into town from the Burro Mountains with a lump of turquoise wrapped in a gunny sack. He offered it for sale and the London agent examined the gem indifferently and finally bought it for a song. He had found his prize at last, and he bore it away to London before the Mexican had time to discover he had thrown away what would have meant to him a fortune. But Mr. Rothchilds had a pair of vases now, and one came from Persia and the other from New Mexico, a perfect match, yet produced at opposite ends of the earth.

Source of Information: Col. W. C. Porterfield, *Kansas City Star*.

Kingston, Ghost Mining Town

by Clay Vaden

Silver Prices Scanned in Town Where Fall Dug Ore

The half dozen families left in Kingston, New Mexico, near "The Devil's Backbone," at the top of the Scenic Black Range Mountains, at the edge of the Gila Forest, hope that the advancing price of silver will restore this ghost mining town to the glory it once enjoyed when Main Street was lined with 22 saloons, one church, and a brewery, and when 1,500 inhabitants produced millions of dollars in ore.

There were schools here in the silver days but Kingston has few families with young children now, as most of its residents are gracefully growing old.

When the daily newspapers reach Kingston two or three days after publication—the town is 30 miles from a railroad—the old timers turn to the market pages for silver quotations, and then scan the other pages for items about Albert B. Fall and until recently for items about his partner and neighbor at one time in Kingston, the late oil magnate, Edward L. Doheny.

Doheny staked out several mining claims in 1880. One of them he named "The Miner's Dream," and it proved to be just that. He borrowed $50 from a saloon keeper and left for California, where fortune smiled on him.

Fall worked in the silver mines at Kingston before he established his large ranch near Three Rivers. He and Doheny began their lifelong friendship in this country.

Sheba Hurst, Mark Twain's humorous character in "Roughing It," was the wit of the town. A plain pine slab bearing the name "Sheba" marks his last resting place in the old graveyard. When Sam Bernard, Mike Moffitt, and Ellsworth Bloodgood tell about the glory that was Kingston's, they always speak of Fall, and the late Doheny and Hurst.

Kingston had its share of gamblers, gunmen, and Indians. The first silver ore was found November 3, 1880, in Ladrone (Thief) Gulch. Development of the mining industry was delayed a year because Chief Victorio and his Apaches took to the warpath and killed scores of prospectors and miners.

Later the camp grew to a town of 1,500. Women and children sought refuge in the three story stone hotel now named "The Victorio," when Indians threatened the community.

Besides its 22 saloons, colorful dance halls, and New Mexico's only woman stage driver, Sadie Orchard, Kingston had a theatre, stores, schools, hotels, and three newspapers. One night it was suggested that Kingston needed a church. Hats were passed. Into them, dance hall girls tossed diamond rings, gamblers dropped stickpins, and miners poured gold nuggets. Reduced to money, the collection totaled $1,500.00 and the church was built. Its walls of stone are still standing. The wall behind the altar bears a unique sign: "THE GOLDEN GATE."

Old Rifle Bears Marks of Man's Fight With Bear

by Clay W. Vaden

Bob Bullward of Santa Rita, New Mexico, has a valuable old rifle on the muzzle of which the marks of a bear's teeth are indented. John Butecke, a pioneer resident of the Gold Dust region of New Mexico, north of Hillsboro, Sierra county, was the man who while living on his mining claim near Gold Dust placer mines told the writer the exciting true story behind the historic gun.

The gun at the time the bear chewed it was held in Butecke's hands. Butecke was following a herd of deer in the Black Range mountains on Bear Creek 14 miles west of Chloride when he came upon a large bear at the top of a bluff. The pathway, he related, was so narrow he could not pass around the sullen bear and after one shot, which wounded the animal, the enraged bear rushed at him and seized the muzzle of the rifle which Butecke had thrust out for protection. A terrific struggle between the bear and man began. The infuriated bear ripped off most of the hunter's clothing, and clamped his teeth over the rifle barrel, holding on with a bulldog tenacity while he (the man antagonist) was trying to point the barrel so he could mortally wound the bear. Finally as Butecke was near exhaustion he succeeded. The bear's hide measured ten feet. Butecke, old timers believe, was one of few New Mexicans who wrestled bare handed with a bear and came out of the struggle alive, though terribly scarred.

Ox Team Freighter Recalls Old Days in Kingston Mine Area

by Clay W. Vaden

"Ox teams were not so fast as the trucks used now to haul ore from the mines hereabout," observes Cebe Goins, pioneer freighter, 90 years young, "but they got the ore out."

Goins drove ten yoke of oxen to freight wagons of seven tons capacity and with tires four inches wide. He later replaced the oxen with 12 teams of mules to each wagon. Goins hauled ore from the paying mines in Kingston district, among them Brush Heap, Gypsy, Blackie, Lady Franklin, Bullion, U.S., Cumberland, Calamity Jane, Keystone, and numbers of others.

When a $1,500 nugget was picked up at Blackie mine, seven miles north of Kingston, a rush to that district followed, he recalled. The Bridal Chamber mine at Lake Valley was one of the best paying in this section of the state. Blocks one yard square of almost pure native silver were often taken from this mine, he said, and it has been roughly estimated that it produced ore worth between five and seven millions of dollars. "There was danger in freighting such rich shipments," said Goins, "and I always had a guard armed with a double barreled shotgun and two six-shooters stay on my wagons until the ore was placed on the cars in Lake Valley."

Goins recalled how the knowledge of ores of Dennis Finley, now a resident of Denver, made him a small fortune.

"A Judge Holt had a lease on and was foreman of the Virginia mine, while Finley was one of the 30 workmen, although he had been foreman of another mine and was a practical mining man," said Goins.

"One day Finley picked up a rich piece of ore and said to Judge Holt, 'This is worth saving.'

"Judge Holt replied, in effect, that if he wanted any information he would ask for it and continued to throw $800 a ton rock over the dump. Finley was given his 'time' in a few days. He obtained a lease from the Virginia Mine company and hauled 18 carloads of high grade ore from the dump. He now owns a chain of stores in Denver but before he made his stake at Kingston he had not seen his family in five years."

Goins came to Sierra county about 1885, living first at Percha, north of Kingston. While several fortunes were taken out of the Kingston mines, he said, the big companies never found official veins, only ores in pockets and chimneys. The Virginia mine is still being worked.

List of Facsimiles

Chilili, Lorin W. Brown, April 8, 1940, NMFWP, WPA #159, NMSRCA ~ 43

Tecolote, Lester Raines, May 13, 1936, NMFWP, WPA #227, NMSRCA ~ 57

The Haunted House, Lester Raines, August 3, 1936, NMFWP, WPA #159, NMSRCA ~ 65

Buried Treasure, Lester Raines, no date, NMFWP, WPA #159, NMSRCA ~ 77

A Discovery of a Cave, Mrs. Frances Totty, November 6, 1937, NMFWP, WPA #159, NMSRCA ~ 101

The Story, F. Totty, September 22, 1937, NMFWP, WPA #159, NMSRCA ~ 105

Lost Treasures, Mrs. W. C. Totty, November 20, 193-, NMFWP, WPA #159, NMSRCA ~ 110

First Mine Registration in New Mexico, 1685, Pedro de Abalos, September 29, 193-, NMFWP, WPA #129, NMSRCA ~ 146

Tin Pan Canyon, Colfax County, Kenneth Fordyce, March 5, 1937, NMFWP, WPA #190a, NMSRCA ~ 158

Glossary of Mining Idioms Used in New Mexcio, by T. F. Bledsoe, June 10, 1937, NMFWP, WPA #161, NMSRCA ~ 174

Letter from Jeff Boone to Mr. Charles Ethridge Minton, September 26, 1940, NMFWP, WPA #184, NMSRCA ~ 200

Dolores, no date, NMFWP, WPA #229, NMSRCA ~ 209

Notes on Pinos Altos Range, Mrs. Mildred Jordan, June 8, 1936, NMFWP, WPA #204, 120 NMSRCA ~ 223

Description of an Old Ghost Town: Georgetown, Mrs. Mildred Jordan, December 7, 1936, NMFWP, WPA #203, NMSRCA ~ 232

List of Illustrations

Photos Courtesy Palace of the Governors
Photo Archives (NMHM/DCA)

COVER: Unidentified Group of Miners, New Mexico 1890?, NMHM/DCA #112985

A. D. Rogers, A. Keith Johnston, Territory of New Mexico, Fray Angélico Chávez History Library, Map Collection (78.9), ca. 1857 ~ 21

According to legend, Indian slave labor was used in the old Spanish mines of New Mexico. Actually, very little mining was done in those days. Simple ladders made of notched logs were used in Indian pueblos and, later, in pit mines. Ritch, 1885, *New Mexico in the 19th Century: A Pictorial History*, Andrew K. Gregg, p. 171 ~ 115

Mexican Arrastra. *Illustrated New Mexico 1885*, W. G. Ritch, Fifth Edition ~ 116

The Placer Mines—working the Rocker, New Mexico (?), Charles F. Lummis, NMHM/DCA #101882 ~ 116

Stoping a mine. This method is not used in a strip mine such as Santa Rita. Thayer, 1888, *New Mexico in the 19th Century: A Pictorial History*, Andrew K. Gregg, p. 172 ~ 117

Lt. Emory's detachment, on reconnoissance, were among the Anglos to pay an official visit to the Santa Rita mines. Emory, 1848, *New Mexico in the 19th Century: A Pictorial History*, Andrew K. Gregg, p. 173 ~ 117

Unidentified miners near Silver City, New Mexico, ca. 1889-92, Rev. Ruben Edward Pierce, NMHM/DCA #93812 ~ 118

Central Mine from the W near Silver City, New Mexico, J. R. Riddle, NMHM/DCA #76118 ~ 118

"U.S. Treasure" mine and mill, Chloride, New Mexico, ca. 1890, Henry A. Schmidt, NMHM/DCA #012633 ~ 119

Dr. Haskell's cabinet of mineral specimens, Chloride, New Mexico, February 17, 1883, Henry A. Schmidt, NMHM/DCA #65570 ~ 120

The town of Lake Valley, west of Hatch, is one of several old mining camps in that area that are now ghost towns. *New Mexico in the 19th Century: A Pictorial History*, Andrew K. Gregg, 133 ~ 120

"Bridal Chamber Mine," Lake Valley, New Mexico, ca. 1890, Henry A. Schmidt, NMHM/DCA #012655 ~ 121

Group of Miners, "Bridal Chamber Mine," Lake Valley, New Mexico, ca. 1890-1895, NMHM/DCA #56218 ~ 121

Lode mining works veins of ore to be processed above ground. Thayer, 1888, *New Mexico in the 19th Century: A Pictorial History*, Andrew K. Gregg, p. 173 ~ 122

Smelting Works at Lake Valley. The Bridal Chamber mine there once produced $3 million in horn silver ore in six months. Thayer, 1888, *New Mexico in the 19th Century: A Pictorial History*, Andrew K. Gregg, p. 133 ~ 122

Stamp Mill, White Oaks or Fort Stanton 1893, Left to Right: 1. Will Lane, 2. I. N. Bailey, 3. Dave Girdwood, 4. Will Real (Lincoln Co.), NMHM/DCA #089680 ~ 123

Miners in Lincoln Country, New Mexico, 1904, NMHM/DCA #005243 ~ 124

"Struck It Big," North Home Stake Mine, White Oaks, New Mexico, NMHM/DCA #163580 ~ 124

Three unidentified miners, Raton/Yankee, New Mexico area, NMHM/DCA #91121 ~ 125

Entrance to coal mine, Raton-Yankee area, New Mexico, 1900?, NMHM/DCA#091138 ~ 125

Raton Coal and Coke Company Coal Mines near Blossburg, NM, ca. 1893, NMHM/DCA #014257 ~ 126

Unidentified Group of Miners, New Mexico 1890?, NMHM/DCA #112985 ~ 126

Pacific Stamp Mill, Pinos Altos, New Mexico, 1890?, L. A. Skelly, NMHM/DCA #148540 ~ 127

Ore was crushed in stamp mills such as this. A drive wheel lifted the vertical hammer rods, letting them fall to pulverize the rock. The thunder of these old mills rumbled across the valleys, and the foundations of these mills still can be seen in abandoned workings. Bell, 1870, *New Mexico in the 19th Century: A Pictorial History*, Andrew K. Gregg, p. 171 ~ 127

Panning gold in the mountains near Cerrillos. Gregg, 1849, *New Mexico in the 19th Century: A Pictorial History*, Andrew K. Gregg, p. 173 ~ 128

Mrs. Captain Jack, Mining Queen of the Rocky Mountains, 1910?, NMHM/DCA #156599 ~ 128

Ruelina Camp and Shaft, Cerrillos, New Mexico, ca. 1881, George C. Bennett, NMHM/DCA #14840 ~ 129

Homestake Shaft, Cerrillos, NM, ca. 1881, George C. Bennett, NMHM/DCA #14843 ~ 129

Entrance to the Grant Tunnel, Chalchihuitl Mine, Cerrilllos, NM, ca. 1881, George C. Bennett, NMHM/DCA #014826 ~ 130

Blowing with Washer, Golden, New Mexico, NMHM/DCA #154787 ~ 130

21 oz. 5 p.n placer Gold. Result of the operations of The Santa Fe Dredging Co., Golden, New Mexico, Parkhurst, NMHM/DCA #12647 ~ 131

Prospector and burros in front of photograph gallery, Hillsboro, Kingston, New Mexico, ca. 1892-1900, George T. Miller, NMHM/DCA #76511 ~ 132

Panning gold near Hillsboro, (?), Davis, Henry A. Schmidt, NMHM/DCA #148168 ~ 132

Man with pickaxe, mining area of Hillsboro and Kingston, New Mexico, 1885-1892?, Henry A. Schmidt, NMHM/DCA #076432 ~ 133

No. 1 Tunnel of Good Hope Mine, Hillsboro, NM, ca. 1890, Henry A. Schmidt, NMHM/DCA #65124 ~ 134

Lady Franklin Mine, Bullion Hill, Kingston, New Mexico, NMHM/DCA #160404 ~ 134

Mrs. Sadie Orchard on right, in front of Ocean Grove Hotel, Hillsboro, New Mexico, 1895-1902, George T. Miller, NMHM/DCA #076560 ~ 135

Chinese laborer, mining area of Hillsboro and Kingston, New Mexico, 1885-1886?, NMHM/DCA #076201 ~ 135

"Elegantly dressed negro woman (possibly one of Sadie Orchard's girls)," mining area of Hillsboro and Kingston, New Mexico, 1885-1892?, J. C. Burge, NMHM/DCA #076431 ~ 136

Malachite Bill, miner, Albright Parlors, Albuquerque, New Mexico, 1890, NMHM/DCA #90392 ~ 137

School of Mines, Socorro, *New Mexico*, Max Frost, 1894 ~ 138

Bibliography of New Mexico Federal Writers' Project Documents

WPA—Works Progress Administration/NMFWP—New Mexico
Federal Writers' Project
NMSRCA—New Mexico State Records Center and Archives

A Discovery of a Cave, Mrs. Frances Totty, November 6, 1937, NMFWP, WPA #159, NMSRCA

A Mine for Two Barrels of Water, W. L. Patterson, no date, NMFWP, WPA #129, NMSRCA

A Prospector's Experience, Mrs. W. C. Totty, July 13, 1937, NMFWP, WPA #129, NMSRCA

An Expedition, Santa Fe Weekly Gazette, February 22, 1868, from New Mexico Federal Writers' Project, no date, NMFWP, WPA #92a, NMSRCA

Buried Money on the Mimbres, Frances E. Totty, September 28, 1938, NMFWP, WPA #159, NMSRCA

Buried Treasure, Genevieve Chapin, May 13, 1936, NMFWP, WPA #159, NMSRCA

Buried Treasure, Lester Raines, no date, NMFWP, WPA #159, NMSRCA

Chilili, Lorin W. Brown, April 8, 1940, NMFWP, WPA #159, NMSRCA

Citizenship Papers: Celestino Carrillo, Ernest Prescott Morey, August 22, 1938, NMFWP, WPA #129, NMSRCA

Coal in Colfax County—New Mexico, Kenneth Fordyce, April 12, 1937, NMFWP, WPA #129, NMSRCA

Description of a Mine: Santa Rita Copper Mine, Mrs. Mildred Jordan, May 29, 1936, NMFWP, WPA #204, NMSRCA

Description of a Point of Interest: Steeple Rock Peak, Mrs. Mildred Jordan, June 8, 193?, NMFWP, WPA #129, NMSRCA

Description of an Old Ghost Town: Georgetown, Mrs. Mildred Jordan, December 7, 1936, NMFWP, WPA #203, NMSRCA

Dolores, no date, NMFWP, WPA #229, NMSRCA

Elizabethtown, Kenneth Fordyce, February 5, 1937, NMFWP, WPA #190a, NMSRCA

First Mine Registration in New Mexico, 1685, Pedro de Abalos, September 29, 193-, NMFWP, WPA #129, NMSRCA

Founding of Silver City, Mrs. W. C. Totty, November 17, 1937, NMFWP, WPA #203, NMSRCA

Gambusinos, Lorin W. Brown, December 5, 1938, NMFWP, WPA #159, NMSRCA

Glossary of Mining Idioms Used in New Mexico, by T. F. Bledsoe, June 10, 1937, NMFWP, WPA #161, NMSRCA

Gold, D. D. Sharp, February 6, 1937, NMFWP, WPA #129, NMSRCA

Gold Fever in Ojo de la Casa as Told by Patricio Gallegos, Stories of William Eckert and Juan Maria Gallegos, J. P. Batchen, no date, NMFWP, WPA #159, NMSRCA

Gold Gulch Findings, Frances E. Totty, May 23, 1938, NMFWP, WPA #129, NMSRCA

Hands That Built America: Memorandum On Mining Operations, from New Mexico Writers' Project, no date, NMFWP, WPA #129, NMSRCA

Hidden Treasures, Mrs. Frances Totty, November 17, 1937, NMFWP, WPA #159, NMSRCA

Humorous Incidents of Early Mining Days: A False Alarm, James A. Burns, July 18, 1936, NMFWP, WPA #129, NMSRCA

Humorous Incidents of Early Mining Days: A Self Made Reputation, James A. Burns, July 18, 1936, NMFWP, WPA #129, NMSRCA

Indian Fight in the Floridas, Betty Reich, March 5, 1937, NMFWP, WPA #86, NMSRCA

Indian Stagecoach Robbery, Betty Reich, July 17, 1937, NMFWP, WPA #213, NMSRCA

Jessie Martin, Desert Rat, N. Howard Thorp, no date, NMFWP, WPA #159, NMSRCA

Kingston, Ghost Mining Town, Clay W. Vaden, April 13, 193?, NMFWP, WPA #230, NMSRCA

La Mina de la Virgen de Oro or The Mine of the Golden Virgin, N. Howard Thorp, January 4, 1937, NMFWP, WPA #129, NMSRCA

La Mina Escondida: The Hidden Mine, N. Howard Thorp, January 28, 1937, NMFWP, WPA #159, NMSRCA

Letter from Jeff Boone to Mr. Charles Ethridge Minton, September 26, 1940, NMFWP, WPA #184, NMSRCA

Lost Mine of the Pedernal, N. Howard Thorp, August 5, 1937, NMFWP, WPA #159, NMSRCA

Lost Treasure, Manuel Berg, no date, NMFWP, WPA #159, NMSRCA

Lost Treasures, Mrs. W. C. Totty, November 20, 193-, NMFWP, WPA #159, NMSRCA

Lost Treasures of Grand Quivira, Edith L. Crawford, July 6, 1937, NMFWP, WPA #159, NMSRCA

Louis Ancheta, Frances E. Totty, September 26, 1938, NMFWP, WPA #203, NMSRCA

Mina de la Tierra, Robert Pfanner, April 4, 1936, NMFWP, WPA #129, NMSRCA

Mines of Northern New Mexico (Historical), James Burns, February 29, 1936, NMFWP, WPA #129, NMSRCA

Mining and Minerals, Harriett Brent, rewritten by Robert Pfanner, edited by Carlotta Warfield, no date, NMFWP, WPA #229, NMSRCA

Mining Life, Frances E. Totty, December 28, 1937, NMFWP, WPA #86, NMSRCA

Mining Stories from Las Placitas: Legend of Montezuma Mine, from the New Mexico Federal Writers' Project, no date, NMFWP, WPA #159, NMSRCA

Negro Findings, Frances E. Totty, May 23, 1938, NMFWP, WPA #159, NMSRCA

Notes on Pinos Altos Range, Mrs. Mildred Jordan, June 8, 1936, NMFWP, WPA #204, NMSRCA

Old Rifle Bears Marks of Man's Fight With Bear, Clay W. Vaden, September 1, 1936, NMFWP, WPA #230, NMSRCA

Out of Bondage (José de Luz Seeks His Fortune), Lou Sage Batchen, May 7, 1941, NMFWP, WPA #224a, NMSRCA

Ox Team Freighter Recalls Old Days in Kingston Mine Area, Clay W. Vaden, October 7, 1936, NMFWP, WPA #230, NMSRCA

Pinos Altos: The Great Gold Producing Camp of the County, Frances E. Totty, May 9, 1938, NMFWP, WPA #129, NMSRCA

Pioneer: Hidden Treasure (Cutter, Sierra County), Lester Raines, March 21, 1936, NMFWP, WPA #159, NMSRCA

Place Names: Cities, Towns, and Villages, Lincoln County, Edith L. Crawford, March 2, 1939, NMFWP, WPA #211, NMSRCA

Prospector 40 Years, Edith L. Crawford, March 10, 1939, NMFWP, WPA #210, NMSRCA

Reminiscence of an Old Prospector, Ernest Prescott Morey, August 15, 1938, NMFWP, WPA #129, NMSRCA

Sad Disaster at Mogollon, Taken from Santa Fe New Mexican, Aug. 20, 1896, John Looney, May 26, 1939, NMFWP, WPA #231, NMSRCA

Secrets of the Guadalupes, Lorin Brown, July 31, 1939, NMFWP, WPA #159, NMSRCA

Silver City Mines, Frances E. Totty, December 28, 1937, NMFWP, WPA #129, NMSRCA

Snakes in a Mine Shaft, Mrs. W. C. Totty, June 19, 1937, NMFWP, WPA #129, NMSRCA

Still Buried Treasure, Hermione Manning, L. Raines, July 25, 1936, NMFWP, WPA #159, NMSRCA

Tecolote, Lester Raines, May 13, 1936, NMFWP, WPA #227, NMSRCA

The Adams Diggings, E. V. Batchler, November 19, 1938, NMFWP, WPA #159, NMSRCA

The Church of the Golden Bell, L. Raines, August 3, 1936, NMFWP, WPA #159, NMSRCA

The Dead Burro Mine, N. Howard Thorp, June 8, 1938, NMFWP, WPA #159, NMSRCA

The Founding of Silver City, Taken from the Mogollon Mines, Published in 1914, Frances E. Totty, September 15, 1938, NMFWP, WPA #202, NMSRCA

The Ghost of Georgetown, Mrs. W. Totty, May 21, 1937, NMFWP, WPA #203, NMSRCA

The Ghost of Priors Canyon, Frances E. Totty, October 20, 1938, NMFWP, WPA #159, NMSRCA

The Haunted House, Lester Raines, August 3, 1936, NMFWP, WPA #159, NMSRCA

The Helen Rae Mine, W. L. Patterson, October 19, 1936, NMFWP, WPA #129, NMSRCA

The Lost Gold Mine, Kenneth Fordyce, March 13, 1937, NMFWP, WPA #159, NMSRCA

The Lost Juan Mondragon Mine, Lorin W. Brown, no date, NMFWP, WPA #159, NMSRCA

The Lost Mine, Betty Reich, June 18, 1937, NMFWP, WPA #196, NMSRCA

The Lost Sublett Mine, Kathryn Ragsdale, June 29, 1936, NMFWP, WPA #159, NMSRCA

The North and South Homestake Mines, Edith L. Crawford, July 24, 1939, NMFWP, WPA #129, NMSRCA

The Old Abe Mine (I), Edith L. Crawford, August 5, 1939, NMFWP, WPA #129, NMSRCA

The Old Abe Mine (II), Edith L. Crawford, August 11, 1939, NMFWP, WPA #129, NMSRCA

The Old Mines Near Las Placitas, N. Howard Thorp, no date, NMFWP, WPA #159, NMSRCA

The Schaeffer Diggings, Betty Reich, February 27, 1937, NMFWP, WPA #159, NMSRCA

The Story, F. Totty, September 22, 1937, NMFWP, WPA #159, NMSRCA

The Story of Adam's Diggings, L. Raines, August 3, 1936, NMFWP, WPA #159, NMSRCA

The Treasure of Punta de Agua by Edna Shaw, L. Raines, August 31, 193?, NMFWP, WPA #159, NMSRCA

Tin Pan Canyon, Colfax County, Kenneth Fordyce, March 5, 1937, NMFWP, WPA #190a, NMSRCA

Treasure, Ramitos Montoya, July 25, 1936, NMFWP, WPA #159, NMSRCA

Turquoise, Mrs. W. C. Totty, June 25, 1937, NMFWP, WPA #129, NMSRCA

Turquoise Mines Near Oro Grande, W. L. Patterson, July 6, 1936, NMFWP, WPA #129, NMSRCA

Tyrone Turquoise Mines, Ernest Prescott Morey, September 23, 1938, NMFWP, WPA #129, NMSRCA

Wilcox Mining Claim, Frances E. Totty, December 18, 1937, NMFWP, WPA #129, NMSRCA

Names Index

Adams Diggings (Adams Diggings Mine, Adam's Diggings), 33-35, 38-39, 73, 255
Adams, Edward (Adam), 33-39, 73
Aklin, Henry, 112
Aklin's store, 112
Alabama Group, 196
Albert Larson and Co., 238
Aldano, Jose, 63
Allen, Col. Richard, 231
Allen, J. M., 153, 156
Allen, John Wessley, 182
Allen, Otho, 182
Alsberg, Henry G., 15
American Mine, 194
American Smelting and Refining Company, 165
Ancheta (family), 18, 102, 107
Ancheta, Juan, 102
Ancheta, Louis, 101-102, 106-107, 254
Anderson, H. E. (Harry E.), 142, 145
Apaches, 33-34, 36, 78, 81-83, 98, 103, 114, 161, 165, 170-171, 178-179, 192, 220, 225, 228, 234-235, 238, 242
Applegate. Frank O., 55, 76
Aranda, Francisco, 202
Aranda, Pablo, 201-203
Aranda, Thomas (Tomas), 201
Archbishop, 171
Archuleta, Onecimo, 108-109
Archuleta, Pedro, 142
Armmene, Adolph, 183-185
Arnold, Messer., 228
Arthur, Lucius, 61

Aul, Duane, 73
Aztec Mine, 148
Azure Mining Co., 237-238
Bailey, N., 123, 249
Banks, L. J., 154
Barachman, John, 108
Barela, Trinidad, 108
Bartolino, Babe, 158
Batchen, J. P., 25, 253
Batchen, Lou Sage, 28, 254
Batchler, E. V., 33, 255
Baxter, George, 151-152
Benigno, 49
Bennett, Col. Joseph, 233-234
Bennett, George C., 129-130, 250
Berg, Manuel, 40, 254
Bernard, Sam, 241
Billy the Kid (William Bonney), 48, 58, 196
Birch, 106, 224
Bisbee, Frank, 233
Black, Honorable Robert, 234
Black, Messer., 234
Blackie Mine, 244
Bledsoe, T. F., 174, 247, 253
Bloodgood, Ellsworth, 241
Bloomer, Mr., 144-145
Boone, Jeff, 200, 247, 254
Borgada, Archbishop, 108
Brakebill, Charles ("Charley"), 221
Bremen, M. W., 228-229
Bremen's mill, 220-221
Brent, Harriet, 198, 254
Bridal Chamber Mine, 18, 121-122, 244

Bronson, Larry, 147
Brown, Lorin W., 17, 43-44, 47, 49, 247, 252-253, 255-256
Brush Heap Mine, 244
Buford, Chas., 156
Bullard, James (James B., "Jim"), 217, 227, 229, 233
Bullard, John, 217, 223, 229
Bullard, Mr., 229
Bullion Mine, 244
Bullward, Bob, 243
Burge, J. C., 136, 251
Burk, Charlie, 213-216
Burnet, R. M., 67,
Burns, James A., 141, 143, 147, 253-254
Burr, Mrs. J. J., 166
Butecke, John, 243
Butler, Sidney, 180
Butterfield stage line, 48
Calamity Jane Mine, 244
Campbell, James S., 217-218, 227
Carasco, Mr., 230
Carley, 201-202
Carlos, 40-42
Carlyle Mine, 167
Carney, Barney, 94,
Carrasco, Lieutenant Colonel, 165
Carrasso, Colonel (Don), 176-178
Carrillo, Celestino (Caelestino Carrillo), 170-173, 182, 252
Carrillo, Clair, 172
Carrillo, Sofa, 172
Carrington, Mr., 142
Cash Entry Mine, 44
Cassidy, Ina Sizer, 15-16
Central Mine, 118
Cerasco, Mr., 227
Chalchihuitl Mine (Chalchihhuetl), 130, 198, 250
Chapin, Genevieve, 54, 252

Chase, 156
Chino Copper Company, 164, 172
Cibola Reduction and Smeltering Works company, 229
Clark, Richard F., 223
Clark, Senator, 59
Close, Selma (Selma Close Roach, Miss, Frauline), 184, 186
Cochise, 82, 178, 181, 225
Cochitis, 211-212
Coleman, Mr., 229
Colonial Mining Company, 168
Colorado, Mangas ("Mano," "Red Hand"), 103, 178, 181
Commercial Mine, 232
Confidence Mine, 168
Connors, A. B., 235
Connors, John, 235
Conway, Jay, 160
Cooney, Captain, 200
Copper Mine, 161
Coronado, 191, 205
Cosgrove, Messer., 234
Cox Ranch, 192
Crawford, Edith L., 18, 56, 58, 150-151, 153, 156, 254-256
Cumberland Mine, 244
Cummings, Billy, 223
Davidson, 33-38
Davis, Captain N. S., 149
Davis, Ed, 94
de Abalos, Antonio, 146
de Abalos, Pedro (Avalos), 146, 247, 253
de Aguilar, Captain Don Alonzo Rael, 146
de Baca, Alvan Nuñez Cabeza, 198
de Cruzate, Don Dominguez Jironza Petris, 146
de Narieza, Justice Garcia, 146
de Niza, Fray Marcos, 187
De Vargas, 56, 187

Dead Burro Mine, 96, 99
Deep Down Mining Company, 168
del Ao., Alploniso, 146
del Campo, Santa Inez, 51
Delgado, Friar, 206
DeMueles, Amos J., 195
DeMueles Mine, 195
Dexter Mine, 229
Doheny, Edward L., 241
Dolge, Lucile (Miss), 185
Eckert, William, 27, 253
Edison, Thomas, 199
Edwards, George, 178-181
Elgua, Manuel, 177-178
Elsmere, Brother, 92, 95
Emory, Lt., 117, 248
Fall, Albert B., 241
Fallon, Thomas ("Tom"), 65, 72
Fechet, Capt., 221
Fergusson, H. B., 153-154, 156
Fields, Albert, 227
Finley, Dennis, 244-245
Fleming, Mrs. Gene, 110
Foley, Tim, 148
Fordyce, Kenneth, 59, 158-159, 161, 247, 252-253, 256
Francisco, Old, 26-27
Frost, Max, 138, 251
Frost, Sergeant Max, 234
Fusion, H. M., 227
Gallardo, 201
Gallegos, Donaciano, 18, 206-208
Gallegos, Juan Maria, 27, 253
Gallegos, Patricio, 25, 253
Gambusinos, 44-46
Garcia (family), 50
Garcia, Don Pedro, 211
Garcia, Teodillia, 170-172
Garrett, Pat, 58, 196
Geronimo, 48, 103, 238

Girdwood, Dave, 123, 249
Glover, Mr., 66
Goforth, Leslie K., 222
Goins, Cebe, 18, 244-245
Gold Coin Mine, 201-202
Gold Standard Mine, 201-202
Gomez, Frank, 92
Good Hope Mine, 134, 250
Gregg, Andrew K., 115, 117, 120, 122, 127-128, 248-250
Gregg, Dr. Josiah, 197
Guerin, Padre, 50,
Gurule, Don Jose, 210
Gurule, José, 32
Gypsy Mine, 244
Hale, Green, 167
Hall, E. F., 141-142, 144-145
Hamilton, Chas., 156
Hanover Mine, 225
Haskell, Dr., 120
Hawkes, John ("Old John"), 69-71
Hayes, M. D., 166
Hearst (estate), 166
Helen Mining Company, 168
Helen Rae Mine, 194, 256
Helman, Corinne (Miss, Frauline), 184
Hernandes, Adolpho, 171
Heron, Gene, 104
Hewitt, John Y., 153-154, 156-157
Hicks, 106, 224
Hockradle, Jerry, 58
Holmes, Arthur, 193
Holt, Judge, 244
Hoover, Herbert, 167
Hopewell, Willard S. ("Colonel"), 93, 95
Houghes, Mike, 107, 112-113
Houghes, Nick, 112
Hoyle, W. M., 154, 156
Hunner, Jon, 17, 20
Huntington Mill, 201-202

Hurst, Sheba, 241
Ibex Mine, 194
Isleta, Primie, 102
J. B. Lippincott Company, 55
J. H., 90
Jaramillo, Jose Maria, 39
Jhu, 103-104
Jimenez, Antonio, 210, 212
Jironca, Domingo, 146
Johnson, E. H. ("Ed"), 141-142, 144-145
Johnson, J. R., 230
Johnson, Mr., 167
Johnson, Mr. (assayer), 227
Johnston, A. Keith, 21, 248
Jones, F. A., 194
Jordan, Mildred, 164, 167, 223, 232, 247, 252, 254
Joseph, Saint, 211
Juan Mondragon Mine, 49, 256
Kay, Eleanor, 55
Keautzy, Joseph, 214
Kelly, 147
Kelly, Captain, 223
Kelly, George, 141-142
Kelsey, O. D., 153
King Wade's Ranch, 233
Kinsinger, Peter, 147
Kirk, Joseph, 227
Knight, John, 168
Knight, Richard S., 233, 235
Knight, Thomas, 168
Kreonig, William, 147
Keystone Mine, 244
La Mina de Coloa, 210
La Mina de Nepomuceno, 210
La Mina de Ventana, 210
La Mina de la Escalera, 210
La Mina de la Terra, 189, 197, 254
La Mina de la Virgen de Oro ("The Mine of the Golden Virgin"), 18, 205-206, 254

La Mina Escondida ("The Hidden Mine"), 88-89, 91, 254
La Real de Dolores, 44
Lacy, Ann, 14, 17, 20
Lady Franklin Mine, 134, 244, 250
Lane, Will, 123, 249
Langston, 224
Larzola, Governor, 108
Lawrey, Joseph, 148
Legara, Francisco Pablo
Lewis, Bob, 34, 39
Lincoln County Light and Power Company, 150
Livingston, A. P. ("Livingstone"), 153, 156
Livingston, Carl B., 48
Llewellyn, W. H., 196
Long, Ed, 48
Looney, John, 168, 255
Los Cerrillos Turquoise Gem Corporation, 198
Lost Mine of the Pedernal, 92, 254
Lost Sublett Mine ("Sublett Mine"), 60, 66, 67, 256
Lost Tayopa Mine, 81
Louis, Old Joe, 179-180
Lovato, Santos, 25
Lucero, Margaret M., 57
Lummis, Charles F., 116, 248
Luna, Juan, 63
Lund, 156
Lynch, Matthew ("Matt"), 148-149
Lyons, Pat, 148
Magruder, John B., 178, 181
Maiad Queen Mine, 232
Malachite Bill (Parlors, Albright), 137, 251
Manning, Hermione, 69, 255
Marston (Brothers), 224
Marston, Captain, 225
Marston, Tom, 225

Martin, Jesse ("Old Jessie"), 84–87, 253
Martinez, Emilia, 90-91
Maxwell, Lucien B., 149
Mayer, Charles D., 150
McCready, Mrs. A. E., 61
McGregor, Alec, 112-113
McKenna, James A., 83
McKinly Mine, 201
McKnight, Robert, 166
Medina, 172
Menchero, Friar Juan Miguel, 206
Miller and Skillecorn's blacksmith shop, 227
Miller, George T., 132, 135, 250
Mine of Quemado, 89
Minter, Virginia, 67
Minton, Charles Ethrige (Ethridge), 16, 200, 247, 254
Moffett, G. E., 193, 196
Moffitt, Mike, 241
Mogollon Mine, 233
Mondragon, Filogonio, 52-53
Mondragon, Juan, 18, 49-53
Mondragon, Tonita, 51
Montezuma, 211
Montezuma Mine, 25-26, 30, 210, 212, 254
Montoya, José de Luz (José de Luz), 28-32, 254
Montoya, Ramitos, 62, 256
Montoya, Salvado, 57
Moore, Elizabeth (Elizabeth Moore Lowrey), 148, 162
Moore, John, 148
Moore, W. H., 147
Moreno Gold Gravel Company, 141, 144
Morey, Ernest Prescott, 170, 176, 183, 252, 255, 257
Morgan, Elizabeth, 65, 72, 74, 77
Morrells, John B., 234

Mountain Key Lode, 225
Mountain Key Mill, 224
Mrs. Captain Jack, 128, 250
Muno, Carlos, 63
Murray, P. H., 203
Murray, W. D., 184
Nannie Baird Mine, 192
Navajos, 64
Navarez, 110
Nepomoseno, 89-91
Nevada Consolidated Copper Co., 164
New Issue Mine, 229
New Mexico, Governor of, 205
New Mexico Mining Co., 209
New Mexico, Surveyor General of, 205
New Spain, Viceroy of, 205
Nick, 237-238
Norero, Carlos, 225
North and South Homestake Mines, 124, 129, 151-152, 256
Nussbaum, Aileen O'Bryan, 16
Ocean Grove Hotel, 135
Old Abe Company, The, 154-155
Old Abe Mine, 153-154, 156-157, 256
Old Glass Mill, The, 154
Onate, 205
Onizz, Juan, 165
Orchard, Sadie (Mrs.), 135, 136, 250-251
Orchard, Sadie, (stage driver), 242
Ortiz (family), 190
Otermin, 187, 206
Otero, Carmelita (Miss), 202
Ownby, Judge B. B., 167
Pablo, 40-42
Pacheco, Isaro, 40, 42
Pacific Stamp Mill, 127
Paiutes, 59
Parker, J. A., 195
Parker, Tom, 183
Patterson (Bros.), 156

Patterson, H. J., 156
Patterson, C. Ewing, 156
Patterson, W. L., 192, 194-195, 252, 256-257
Pattie (family), 165-166
Pattie, James, 165-166
Pedro, 207-208
Perea, Don José Leander, 29
Perea, Francisco ("Colonel"), 85
Perez, Bialkin, 63
Perkins, S. M. ("Ole Perk"), 192-193
Pfanner, Robert, 197, 199, 254
Phelps-Dodge Corporation, 160
Pierce, Rev. Ruben Edward, 118, 248
Pike, Lieutenant, 165, 181
Pino, Friar, 206
Pope Mill, 230
Porterfield (brothers), 238
Porterfield Turquoise Mines Co., 237-238
Porterfield, W. C. (Col.), 237, 240
Poseyemo, 211
Prichard, Colonel, 156
Providence Mine, 230
Pueblos, 183, 187
Quemadenos, 89
Queres, 198
Quien Sabe Mine, 232
Rae, John, 194
Rael-Gálvez, Estévan, 13
Raesequeon, Manuel, 101
Raesequeon, Pedro, 18, 101
Ragsdale, Katherine, 66, 256
Raines, Lester, 57, 65, 68, 75, 247, 252, 255
Raines, L., 69, 72-73, 77, 255-256
Raton Coal and Coke Company Coal Mines, 126
Raton Coal and Coking Company, 159
Real, Will, 123, 249
Red Eye, 65
Redman, 156
Reeds, 234
Reich, Betty, 23, 78, 80-82, 253, 256
Reyes, Juan, 96-99
Riddle, J. R., 118, 248
Ripley, Mr. E. P., 159
Ritch, W. G., 116, 248
Roach, William, 186
Robert E. Lee Lode, 153
Rockford Mine, 194
Rodgers, Clark, 100
Rodgers, Roscoe, 100
Rogers, A. D., 21, 248
Romero, Donaciano, 49-51
Romero, Eulalia, 50
Romero workings, 176
Roosevelt, Franklin Delano, 15-16
Rose, A. B., 194
Ross, H. L., 155
Ross, Mr., 68
Rothchild (family), 183
Rothchild, Louis, 183-185
Rothchilds, Mr. (of London), 239-240
Rothchilds of Maiden Lane, 238
Rubio, Juan, 63
Rymer, Elise, 11
Rynerson, Col., 229
Saiz, (family), 97
Saiz, Don Antonio, 97
Sager, Mr. Frank J., 154
Samora, Felix, 212
San Jose Mine, 225
Sanborn, Captain, 37
Sanderson, John, 213
Sandias, 31
Santa Fe Dredging Co., 131
Santa Rita Copper Mine, 164-166, 173, 177-178, 181, 225
Sarafin, 206-207
Schaeffer Diggings, 82, 256

Schaeffer, Jake, 82-83
Schmidt, Henry A., 119-121, 132-134, 248-250
Scott, O. L., 220, 230
Seneca Mine, 229
Senora del Pilar de Zaragosa (mine), 146
Sewell, G. A., 198
Sharp, D. D., 201, 253
Shaw, Edna, 75, 256
Sibley, General, 166
Sigafus, Jas. M., 151-152
Silver Bell Mine, 232
Simpson, Messer., 228
Singleton, 98
Siquieros, Leonardo, 224
Skelly, L. A., 127, 249
Skillicorns' Mill, 225
Slough, Dr., 216
Snively, 106, 224
Snyder, Agnes Meader, 114, 221
Sowell, Erma, 68,
Spain, King of, 88
Speigelberg-Hockradle, 156
St. Louis, Rocky Mountain & Pacific Company, 159
Stamp Mill, 123
Stapp, N. H., 57
Stevens, Bob, 112
Stevens, Robert, 102, 106
Sublett, Ben ("Old Ben," "Old Sublette"), 47-48, 60-61, 66
Sublett, Ross, 61, 66
Summer, Peter, 214
Sumner, Colonel, 178
Swartzs, Jesse, 173
Swartzs, Jim, 173
Swilling, Lieutenant, 225
Swishelm, John
T. & P. Railroad, 47
Tabor, H. A. A., 203

Taylor, General, 179
Taylor, Jim, 223
Tennessee Reduction Company, 230
Terrazas, 78
Tezcuco, 236
Thompson, Mrs. Alex, 80
Thorp, N. Howard, 18, 84, 88, 92, 96, 205, 210, 253-256
Tidwell, Messer., 228
Tiffany and Company, 184-185, 198
Tiffany Mine, 196
Tinaja, Estaban Back, 77
Torrero, Giorge, 98-99
Torrero, Manuel, 96-99
Torrez, Hijinie, 51
Totty, Frances E. (Mrs., Mrs. W., Mrs. W. C.), 18, 100, 103, 105-106, 108, 110-111, 114, 139, 213, 217, 219-220, 222, 224, 227, 231, 233, 236, 247, 252-257
Trall, Miss, 109
Twain, Mark, 241
Twitchell, Ralph Emerson, 43, 146, 165, 196
Two Ikes Mine, 229-230
Tyrone Turquoise Mines, 183, 185, 237, 257
Utes, 161
Utter, George, 167
Vaden, Clay W., 18, 241, 243-244, 253-255
Valley-Fox, Anne, 14, 17, 20
Van Potten, Colonel, 103
Vargas, Pasquel, 176
Vedders, Lieutenant, 234
Vera Cruz Mine, 194
Victorio, 48, 78-79, 82, 242
Virgin Mary, 212
Virginia Mine, 244
Virginia Mine Company, 245
Vogel, Alfonse, 185
Vogel, Felix, 183-185

Wall, John, 194
Wallace, Lew (General, Governor), 85-86
Ward, Ed, 112
Warfield, Carlotta, 199, 254
Watson, Wm., 153-154, 156
Webb, Messer., 228
Webster, Mr., 227
Weldon, Frank, 167
White, 54-55
White Oaks Lode, 153
Whitehill, Messer., 228
Whitmey, J. Parker, 166
Wilcox Mining Claim, 114, 257
Wilcox, Mr., 114
Williams, Tom, 150
Wilson, 26
Wilson (young man), 212
Wilson, John E., 151-152
Wilson, Mike, 61
Winters, Jack, 151-152
Wisconsin Mining Company, 229
Wood, General, 179
Wooton, Dick, 25
Wooton Mine, The, 25
Yankie, Joseph, 227, 229
Ybarra, 110
Yellow Pole Ranch, 201
Young, 82
Yrisarri, Jacobo, 187
Zarate-Salmeron, Father, 198, 206

www.ingramcontent.com/pod-product-compliance
Lightning Source LLC
Chambersburg PA
CBHW020834160426
43192CB00007B/645